PLUGGED IN

TECHNOLOGY, RHETORIC, AND CULTURE IN A POSTHUMAN AGE

NEW DIMENSIONS IN COMPUTERS AND COMPOSITION
Gail E. Hawisher and Cynthia L. Selfe, editors

Digital Youth: Emerging Literacies on the World Wide Web
Jonathan Alexander

Role Play: Distance Learning and the Teaching of Writing
Jonathan Alexander and *Marcia Dickson* (eds.)

Aging Literacies: Training and Development Challenges for Faculty
Angela Crow

Writing Community Change: Designing Technologies for Citizen Action
Jeffrey T. Grabill

Literacies, Experiences and Technologies: Reflective Practices of an Alien Researcher
Sibylle Gruber

Datacloud: Toward a New Theory of Online Work
Johndan Johnson-Eilola

Digital Writing Research: Technologies, Methodologies and Ethical Issues
Heidi A. McKee and *Danielle DeVoss* (eds.)

At Play in the Fields of Writing: A Serio-Ludic Rhetoric
Albert Rouzie

Integrating Hypertextual Subjects: Computers, Composition, Critical Literacy and Academic Labor
Robert Samuels

Multimodal Composition Resources for Teachers
Cynthia L. Selfe (ed.)

Sustainable Computer Environments: Cultures of Support in English Studies and Language Arts
Richard Selfe

Doing Literacy Online: Teaching, Learning, and Playing in an Electronic World
Ilana Snyder and *Catherine Beavis* (eds.)

Labor, Writing Technologies and the Shaping of Composition in the Academy
Pamela Takayoshi and *Patricia Sullivan* (eds.)

forthcoming:

Webbing Cyberfeminine Practice: Communities, Pedagogy and Social Action
Kristine Blair, Radhika Gajjala, and *Christine Tully* (eds.)

PLUGGED IN

TECHNOLOGY, RHETORIC, AND CULTURE IN A POSTHUMAN AGE

edited by

Lynn Worsham
and
Gary A. Olson

Foreword by
Cynthia Y. Selfe

HAMPTON PRESS, INC.
CRESSKILL, NJ 07626

Copyright © 2008 by Hampton Press, Inc.

All rights reserved. No part of this publication may be transmitted or reproduced in any media or form, including electronic, mechanical, photocopy, recording, or informational storage and retrieval systems, without the express written consent of the publisher.

Printed in the United States of America

Library of Congress Cataloging-in-Publication Data

Plugged in : technology, rhetoric, and culture in a posthuman age / edited by Lynn Worsham and Gary A. Olson ; foreword by Cynthia Y. Selfe.
    p. cm. — (New dimensions in computers and composition)
Includes bibliographical references and index.
    ISBN-13: 978-1-57273-833-1 (hbk.)
    ISBN-13: 978-1-57273-834-8 (pbk.)
1. Rhetoric—Social aspects. 2. Mass media and technology. I. Worsham, Lynn, date II. Olson, Gary A., date
    P301.5.S63P58 2008
    808'.0420285—dc22
                      2007052776

Printed and bound in the United States of America on acid-free paper that meets the standards of the American National Standard for Permanence of Paper for Printed Library Materials

        10   9   8   7   6   5   4   3   2   1

Hampton Press, Inc.
23 Broadway
Cresskill, NJ 07626

CONTENTS

Foreword
    Cindy Selfe vii

Introduction
    Lynn Worsham and Gary A. Olson ix

PART 1: Cyberbodies in a Posthuman Age

Cybernetics, *Ethos*, and Ethics:
The Plight of the Bread-and-Butter-Fly
Kristie S. Fleckenstein 3

Screening (In)Formation:
Bodies and Writing in Network Culture
Jennifer L. Bay 25

Our Cyberbodies, Ourselves: Conceptual Grounds
for Teaching Commodities to Write
Stanley D. Harrison 41

PART 2: Rhetoric and Writing in a Digital Age

The Available Means of Persuasion: Mapping a Theory
and Pedagogy of Multimodal Public Rhetoric
David M. Sheridan, Jim Ridolfo, and Anthony J. Michel 61

The Political Economy of Computers and Composition:
"Democracy Hope" in the Era of Globalization
M.J. Braun 95

Circuitous Subjects in Their Time Maps
James J. Sosnoski and Ken S. McAllister *125*

Toward a Rhetoric of Network (Media) Culture:
Notes on Polarities and Potentiality
Byron Hawk... *145*

PART 3: Network Culture

Grrrl Zine Networks: Re-Composing Spaces
of Authority, Gender, and Culture
Michelle Comstock *165*

Cyber-Spaces of Grief: Online Memorials
and the Columbine High School Shootings
Maya Socolovsky..................................... *187*

Network Theory and Life on the Internet
John Johnston *207*

Contributors 223

Index... 225

FOREWORD

Cynthia L. Selfe

Most people in our profession recognize that faculty who teach writing in digital environments—in distance education classes, in computer-based labs, in smart classrooms—spend a great deal of time focusing on the tasks of designing challenging and engaging instruction for the digital spaces that many in our profession consider distracting at best and insidious at worst; balancing instruction that effectively blends rhetoric, English studies, and computer applications; figuring out the complicated logistics involved in working with the digital networks, systems, and software that form the radically different digital topographies of our various institutions in the first decade of the twenty-first century.

Importantly, however, and sometimes to the surprise of some in our profession I suspect, the best teachers also spend time *theorizing* technology—although many must struggle to carve out adequate time to do so. The intellectual importance of thinking in theoretical ways about technology cannot be over estimated, any more than can be the intellectual importance of thinking in theoretical ways about teaching and learning and composition and rhetoric. And the importance of theorizing technology can never successfully be separated from teaching with technology.

To a great extent, this is so because theoretical understandings of technology, and about our intimate relationship to it, increase the explanatory power we bring to teaching with digital technologies and within digital environments. Without such understandings—as we have realized—the teaching of English, rhetoric, or composition can become naïve and misdirected. We can make the mistake of forgetting that we or students in our classes are always and already deeply immersed in, fundamentally shaped by, the technology as a complexly related set of cultural formations. We can forget that our own identities as human beings both shape, and are shaped by technology. We can come to ignore the fact that the dimensional relationship between technology and discourse is mutually and continually defined, right in our own classrooms, our own writing, and the exchanges of students. Perhaps most importantly, like some of our colleagues, we can even forget that we are always and already implicated in a profession that depends on and invest in technologies—that paper and printing presses, books and pens, typewriters and fax machines, have shape our discursive work and textual scholarship, our economic and epistemological worldviews, just as intimately, and with as much reach, as computers and digital technologies.

For these reasons, we are proud to publish this collection, edited by Gary Olson and Lynn Worsham, in the New Dimensions series. These ten carefully selected articles from past issues of *JAC* remind us all of the importance, the power, of thinking with theory. Each essay offers insightful and cogent perspectives about teaching, learning, or communicating with technology. And each offers us valuable, theoretically informed ways of understanding ourselves, as teachers and students, as intellectual and social beings, as the humans and cyborgs that we all are and have always been.

INTRODUCTION

Lynn Worsham
Gary A. Olson

Plugged In: Technology, Rhetoric, and Culture in a Posthuman Age is a collection of ten essays examining the role of technology in redefining rhetoric, writing, and "the human." First appearing in *JAC*—the premier journal of theoretical scholarship on the intersections of rhetoric, writing, culture, and politics—these provocative and cutting-edge works explore a range of subjects of keen interest to teachers and scholars of rhetoric and composition.

Part 1, Cyberbodies in a Posthuman Age, examines how technology is inalterably redefining what it means to be a human being. Kristie Fleckenstein employs the metaphor of the boundary-blurring bread-and-butter-fly from C.S. Lewis to theorize a "cyberethos," a conceptual framework for guiding ethical behavior in an era that tends to destabilize ethics. Jennifer Bay theorizes a fresh new vision of an embodied rhetoric by analyzing how the relationship of the writer to the writer's body within networked culture affects the rhetorical situation, a vision that is based on "the notion that the body and computer together form a complex, co-adaptive system." Stanley Harrison also observes how bodies and technology have merged, pointing out that contemporary students have become living commodities; he proposes that composition teachers assist these "cyborg writers" by helping them confront their status as consumers. All three authors are concerned with how to live in this "posthuman" era, a time in which the boundaries between humans and their world are forever blurring.

Part 2, Rhetoric and Writing in a Digital Age, explores the effects of technology on the workings of pedagogy, rhetoric, and public discourse. David Sheridan, Jim Ridolfo, and Anthony Michel analyze how technology has radically transformed the public sphere, deliberative rhetoric, and rhetorical pedagogy, and they map out a reconceptualization of all three within a multimodal context. Rejecting the simplistic opposition presented in some composition scholarship between luddities and cyborgs, M.J. Braun demonstrates how the work of a number of eminent scholars escapes this simplistic opposition, and she argues that scholarship on computers and composition would be further enriched by a more rigorous Marxist critique of capital. James Sosnoski and Ken McAllister discuss how "electronic educational environments" can interpellate students into certain ideological orientations, "subjugating" them to particular ways of reading the world, and they recommend preventing for-profit business from achieving a monopoly on the programming and production of EEEs. Finally, Byron Hawk examines a number of traditional rhetorical concepts in light of the work of philosopher Mark Taylor in order to construct a rhetorical theory based on the concepts of complexity and networks.

Part 3, Network Culture, addresses some of the ways that technology is redefining culture. Arguing that composionists need to take "zine culture" seriously if they wish to continue to situate "academic literacies within the larger framework of cultural production," Michelle Comstock shows how zines produced by members of the grrl subculture have created new forums for "postfeminist authorship." Working from a post-9/11 context, Maya Socolovsky discusses the kinds of work that virtual memorials on the World Wide Web perform, contrasting these sites with physical memorials as sites of memory and mourning. And John Johnston examines the phenomena of "small world networks" and their importance for large dynamic systems, including biological environments.

The ten essays in this book are substantive contributions to the ever-growing scholarly conversations about the role of technology in our professional and personal lives. We hope you find them as stimulating and provocative as we do.

For Leigh

and Riley

PART ONE

Cyberbodies in a Posthuman Age

1
CYBERNETICS, *ETHOS*, AND ETHICS

THE PLIGHT OF THE BREAD-AND-BUTTER-FLY

Kristie S. Fleckenstein

> "Crawling at your feet," said the Gnat (Alice drew her feet back in some alarm), "you may observe a Bread-and-butter-fly. Its wings are thin slices of bread-and-butter, its body is a crust, and its head is a lump of sugar."
> "And what does *it* live on?"
> "Weak tea with cream in it."
> A new difficulty came into Alice's head. "Supposing it couldn't find any?" she suggested.
> "Then it would die, of course."
> "But that must happen very often," Alice remarked thoughtfully.
> "It always happens," said the Gnat.
> —Lewis Carroll

The dilemma of Lewis Caroll's bread-and-butter-fly is an apt metaphor for this cybernetic age. When eating to sustain its life, a bread-and-butter-fly dips its sugar cube head into weak tea and cream only to dissolve and become part of that which it eats. To survive, it blurs its own boundaries, which means that its identity as a bread-and-butter-fly disintegrates. We, too, are bread-and-butter-flies; we live as and amid boundaries that materialize, shift, and disappear, only to rematerialize in new forms. Donna Haraway offers the figure of the vampire as a trope for this troubling time in which categorical clarity and sharply demarcated lines of origin are "polluted" (*Modest* 214–17). Pierre Lévy, drawing on Gilles Deleuze, refers to our blurring borders as the deterritorialization of subjectivity and reality, rendering both nomadic and tangential (29).

Johndan Johnson-Eilola claims that we live "in a cloud of data—the *datacloud*—a shifting and only slightly contingently structured information space" that is both symbolic and material (4). So we experience this phenomenon of shifting borders in our textual and corporeal lives. Textually, the pervasive track changes option in Microsoft Word and the quotidian interactivity of cyberspace disrupt our tidy categories of ownership and identity. Novels composed in various hypertextual formats, such as Michael Joyce and J. David Bolter's Storyspace, or hypermedial poetry offered through John Cayley's *Indra's Net*, destabilize the traditional separation of reading and writing, reader and writer. Through the click of a mouse, readers can re/write a poem, a story, or a branch of a story, confusing any sense we might have of original text, author, and reader. Going one step further, the recent phenomenon Gaia Online, which bills itself as the newest anime role-playing community, offers synchronous and asynchronous narrative adventures that unfold as participants jointly create characters and plot points. Coupled with private messaging, gaming, and avatars, Gaia Online dissolves in multiple ways the boundaries among contributors and among reading-writing, speaking-texting identities. What evolves is neither writer nor reader, but "wreader," an identity that renders permeable the previously rigid identities established by the perfect binding and material weight of a codex book (Ryan 12).

Nor is the bread-and-butter-fly phenomenon limited to textual experiences supported by digital technologies, either stand alone or linked in an array of networked communication nodes. The phenomenon of boundary confusion is manifested in our corporeal lives as well. My mother's recent bout with pneumonia resulted in her temporary alliance with an oxygen tank that became an integral part of her physical existence. Her body boundaries spun out into clear, plastic tubes, linking her like a snaky umbilicus to a life-giving oxygen tank on wheels. Lifestyle and family interactions stuttered, stopped, and reconfigured themselves to accommodate my mother's new body, new identity. Similarly, a niece's long-desired pregnancy blurred the boundaries of her body identity, for she could no longer easily separate her subjectivity from that of her developing child. My own children marvel at ultrasound pictures tracking their gestation, tracing the extent to which they and I were at one time indistinguishable. At fifteen, my older daughter now struggles to institute in stark terms just such a separation, realizing, even as she does so, the degree to which we are intertwined. On a larger, more frightening canvas, we see the uncertainty of shifting and dispersed boundaries enacted in the current conflict in and occupation of Iraq, where "friendlies" and "hostiles" cannot be ascertained with any accuracy or permanency. Experts suggest that veterans returning from active duty in Iraq will contend with the lingering, psychological effects of those blurred boundaries. The dilemma of the bread-and-butter-fly is not just an episode in an amusing children's story; blurred boundaries are not merely interesting bits of intellectual play. They are not limited to our interactions with digital or analog technologies. Instead, the reality of the bread-and-butter-fly is life and death, fact and fiction, truth and lie.

Characterized by slippage and dispersal, the poetic, rhetorical, and material realities of this cybernetic era refuse to reside within discrete, inviolable borders. No realm of life is exempt from this boundary confusion. Johnson-Eilola argues that the changes associated with datacloud living "take hold" in the "real world—on the street, in the workplace, and in the home," permeating all aspects of existence (3). As these realities shift and reconfigure, they raise a host of philosophical questions and dilemmas. Preeminent among them is the status of good character and ethicality. If we have no stable boundaries, no stable reality, and no stable subject, how do we judge whose "voice," as well as whose reality, resonates with the greatest ethical authority, the greatest "good character"? In a reality founded on shifting sand, on what rock do we build our belief, our life choices, and our ethical actions?

In this essay, I argue that we can fruitfully address questions of good character and ethicality in this bread-and-butter-fly age through an amalgamation of Aristotelian *ethos* and Batesonian cybernetics.[1] Viewing *ethos* through the lens of cybernetics offers the concept of *cyberethos*, which provides a robust framework for addressing good character and determining ethical behavior in an era that destabilizes both. Through cyberethos, we recognize the fluidity of identity while simultaneously committing ourselves to an ecological ethics. That ethical stance requires us to take responsibility for the embeddedness of good character within a material context permeated by nonlinear time. Cyberethos calls us to act on and judge our inescapable dispersal across osmotic rhetorical and material borders because good character and virtuous behavior are mutually linked.[2] I begin my construction of cyberethos with Bateson's concept of a living system, rereading Aristotle's *ethos* as a "difference that makes a difference." I then turn to the material constraints of Bateson's cybernetics, highlighting the permeability of life lived and life spoken in Aristotle's *ethos*. I next map classical *ethos* according to cybernetics's nonlinear time, arguing for the importance of nonlinear cause and effect in cyberethos. Finally, I conclude with the ethical stance that evolves out of this cybernetic reading of *ethos*, illustrating how cyberethos provides guidelines for an ecological ethics in a bread-and-butter-fly world.

BREAD, SUGAR, AND WEAK TEA
ETHOS AS A LIVING SYSTEM

Fused with cybernetics, Aristotle's account of persuasion by good character offers fruitful ground for theorizing issues of identity and ethical behavior in a cybernetic era.[3] As Aristotle says, via George Kennedy, *ethos* is persuasion through good character and occurs "whenever speech is spoken in such a way as to make the speaker worthy of credence." "[T]he controlling factor in persuasion," *ethos* resides in spoken speech, which, at this point of the *Rhetoric*, refers to thought and content, not style and delivery (1.2.4). However, in spite of this seemingly straightforward account, *ethos* defies Aristotle's struggles to taxonomize it neatly. In an adumbration of bread-and-butter-fly realities and subjectivities, Aristotle's *ethos* morphs across borders, resisting all efforts to hold it stable. In this section,

I explore the ways in which *ethos* can be interpreted as an information system, a living network consisting of rhetor, text, audience, and context. It functions as a difference that makes a difference and offers insights into controversies swirling around online identities.

Cybernetics,[4] a term reclaimed in 1948 by MIT mathematician Norbert Wiener to describe a new probabilistic theory of messages, is a science devoted to investigating questions of control and communication via a theory of messages:

> It is the purpose of Cybernetics to develop a language and techniques that will enable us indeed to attack the problem of control and communication in general, but also to find the proper repertory of ideas and techniques to classify their particular manifestation under certain concepts. (*Human* 17)

Society can only be understood by a study of messages between "man and machine, between machines and man, and between machine and machine" (16). Simply defined, cybernetics is the theory that any entity—machine, human being, society—is constituted and reconstituted by the flow of information throughout transacting and circular pathways by which messages communicate content about the system and ambient environments. An entity exists only as long as those pathways exist. However, neither pathways nor information exists independent of or prior to the other. Weiner humorously rephrases the hen-egg debate to characterize this circularity. Neither the hen nor the egg comes first, Weiner says. Instead, the hen is the egg's way of creating another egg (*God* 35). System and pathways are mutually constitutive, both cause and effect at the same time.

This mutuality depends on a concept of information as "news of difference." An early contributor to the post-World War II Macy Conferences on cybernetics, Gregory Bateson defines information as "the differences that make a difference" (*Steps* 459). We live awash by an infinite number of potential "facts" that are impossible to "know" in toto. Therefore, we engage in a filtering and a selection process. What gets through the filter(s) is difference: the differences that have value for the particular system in which we are immersed. To illustrate, the Trobriand Islanders in New Guinea, Bronislaw Malinowski tells us, do not see resemblances between mother and child or between siblings. Children resemble fathers, and two brothers can resemble the father, but not each other. Similarities between mother and children, perceived by outsiders, do not constitute differences that make a difference in Islanders' system (Bolles 107). But these resemblances (significant aspects of an unknowable reality) do not exist prior to the kinship system within which they are significant. Nor does the kinship system exist outside of the resemblances. The two create each other. Any system comes to be, becomes a something, by means of the valuation of differences because a system is an array of relationships that exists through those differences. Therefore, it is impossible to define "system" before defining "difference"; but "difference" cannot be defined outside of its "system." It is in and through difference that a system evolves, and that system continues as long as the valuation of difference remains credible.

Considering *ethos* through this cybernetic lens reveals that good character also evolves through an analogous circular process, as its constitution through prudence, virtue, and good will indicates. Aristotle tells us: "There are three reasons why speakers themselves are persuasive, for there are three things that we trust other than logical demonstrations. These are practical wisdom [*phronesis*] and virtue [*arete*] and good will [*eunoia*]" (2.1.5). Prudence, virtue, and good will serve as the significant differences by which *ethos* can exist. But, paradoxically, these elements cannot exist prior to the instantiation of *ethos*, which is necessary for prudence, virtue, and good will to be identified as significant. Thus, we have the circularity of cybernetics: *ethos* is dependent on a certain configuration of *phronesis*, *arete*, and *eunoia*, while that configuration is dependent on *ethos*. The manifestation of loops (prudence, virtue, and good will) and the creation of the system (*ethos*) are simultaneously intertwined processes.

To illustrate, in his cybernetic account of genetics, Bateson argues that "[t]he tissues of a plant could not 'read' the genotypic instructions carried in chromosomes of every cell unless cell and tissue exist, at that given moment, in a contextual structure" (*Steps* 153–56). So, too, with *ethos*. Members of a discourse situation cannot "read" each other, cannot regard each other as prudent, virtuous, and of good will, unless they exist at that moment as part of the same *ethos*. Thus, *ethos* does not construct itself in response to the dictates of a solitary displaced actor. No single element of a rhetorical act composes itself autonomously. Instead, it evolves on the basis of the flow of information, enabling rhetor, audience, place, and language to create each other mutually through the establishment of relationships called prudence, virtue, and good will, adapting to one another as a means of maintaining the constancy of those relationships. They render each other and the place they inhabit persuasive. Then they negotiate pathways of information so that the good character of *ethos* is maintained. As both cause and effect, *ethos* cannot be reduced to an essence or located in a single facet of any rhetorical act.

A cybernetic lens reveals the degree to which the circularity of classical *ethos* is reliant on subsystems. Cybernetic systems, even those of single-celled organisms, include subsystems, relationships nestled within relationships. The subsystems consist of an interconnected network of feedback and feedforward loops so that information in one circuit is always about information from another circuit. For example, consider the tangle of subsystems in a simple act of word processing. Meaning spirals through transistors, binary machine language, alphanumeric assembly code, programming language, finally manifesting itself in the icon of the flashing cursor. The words and images that materialize on the screen are statements about the loops nested invisibly within them, subject to the discontinuities permeating the languages and the pathways by means of which those languages constitute themselves. The differences that matter at one level become messages within the system (that is, create new relationships) at a different level, feeding into the system to change it or reconfirm it, maintaining the system by the process of its own self-constitution.

This aspect of cybernetics highlights the fact that good character results from an ensemble performance. It exists in and through the flow of messages, a flow that is the system and constitutes the system simultaneously. We cannot point to any one part of the array of pathways and say, "here lies good character," because it is a property of the entire configuration, situated only in its own flow and traceable back to no single causal force. Slippages in good character can be traced back to similar recursive connections among its constitutive subsystems. The adaptation of a speech to the mood and the character of the audience demonstrates the importance of overlapping loops in classical *ethos*. The essential thrust of Book 2 of the *Rhetoric*, Kennedy claims, is the need of the rhetor to perceive and conform to an audience's *ethos*. Quoting William Grimaldi, Kennedy says that Aristotle shows "the speaker how his ethos must attend and adjust to the *ethos* of varied types of auditors if he is to address them successfully" (164), highlighting the feedback and feedforward of rhetor's and audience's *ethoi*. Kennedy elliptically acknowledges the mobius quality of this aspect of *ethos* in a footnote where he quotes Martin Heidegger. Heidegger argues that Aristotle's analysis of emotion in the second book of the *Rhetoric* constitutes the first systematic treatment of "the everydayness of Being with one another" (124). Publicness, Heidegger explains, has its own way of Being which belongs to a "they," but also requires "moods and 'makes' them for itself" (qtd. in Kennedy 124). The rhetor, then, must do two things. She must consider the moods she needs to create, and she must consider the moods out of which she creates: "It is into such a mood and out of such a mood that the orator speaks. He must understand the possibilities of moods in order to rouse them and guide them aright" (Heidegger, qtd. in Kennedy 124). Pathways overlap as they create themselves, feed differences into other pathways, and then become differences about those differences, feeding back into those initial pathways.

This cybernetic reading of *ethos* offers us cyberethos, and cyberethos highlights three qualities. First, it underscores the inseparability of the *pisteis*. Although Aristotle characterizes *ethos*, *logos*, and *pathos* as separate modes of persuasion, each *pistis* is implicated in the establishment of the other. Thus, cyberethos cannot be disconnected from *logos* and *pathos*. Second, the permeability of subsystems means that it is not just the good character of a thought, an utterance, an individual, a culture, or a species that is implicated in cyberethos; it is the good character of the entire array of pathways in action. C.D.C. Reeve neatly captures these snarled recursive relationships when he notes that, in fifth century BCE Athens, the soul of the *politikos* is indivisible from the soul of the city state (203).

Our inability to locate cyberethos in a single aspect but only in the system provides a valuable frame for investigating a parallel slippage in good character within our bread-and-butter-fly age, especially in today's electronic networks. For instance, one of the unique and problematic products of cyberspace has been the evolution of the avatar, an online identity that a user creates to participate in a computer-generated and computer-mediated world. Ranging from discourse-based MOOs (an object oriented multi-user domain, or MUD) to three-dimensional virtual reality interfaces, computer-generated participant worlds require users to craft an identity or a character that then interacts within a prescribed environment according to prescribed rules. As an identity inhabiting a virtual realm, the avatar is the "site where a user brings, modifies, problematizes, and constructs a sense of self distinct from others" (Gromala 216). Yet, because an avatar is a graphic-discursive representation, it, like cyberethos, is simultaneously an "I" and a "not I," where the separation between "copy" and "original," rhetor and citizen, blur. Diana Gromala notes that, through avatars, "human agency is projected and reflected back to influence the user's subjectivity" (216). Such blurring leads Alvy Ray Smith, cofounder of Pixar Animation, to urge actors, seemingly threatened by their unauthorized representation in and through digital technologies, to get an agent to protect "their" avatars, their digitized presence. By explicitly fusing virtual reality (VR) and real world (RW) identities into "animator-actor" and then by ignoring the movement across and the implications of the hyphen, Smith morphs questions of agency and identity into questions of commerce. The key issue for animator-actors will be one of money. Who will command the greater salary: the avatar or the actor (78)?

The constitution of cyberethos as the flow of information along an array of pathways invites us to reconceive cyberspace identity as similarly crafted out of impinging loops of significant differences. From this perspective, we cannot so cavalierly reduce human and avatar identity to intellectual property and contractual commercial negotiations, complex as those issues might be. Because subsystems are lodged within and constitute a larger system, we can only artificially separate life-on-the-screen from life-before-the-screen, the life-of-the-rhetor from the speech-of-the-rhetor. Treating online activities as if they were merely words, pixilated marks on a screen, erroneously isolates one loop of this complex array, leading to the temptation to act without accountability. The seduction of cyberspace, Sherry Turkle notes, is the lure of acting irresponsibly. If there is no investment, no permanence—merely transient marks on a screen—then there is no need to attend to one's reputation (see also McDonough 865). However, cyberethos as a living system reveals the falsity of this isolation. What occurs on the screen cycles back to what happens in front of the screen in the feedback/feedforward phenomenon of cyberethos, tying the soul of the user to the soul of the avatar.

As a permeable ecology of information pathways, a cybernetic *ethos*, a cyberethos, emphasizes that a change in one loop of an ecology cannot be restricted to that single pathway. Rather, it is dispersed throughout the ambient network. In addition, cyberethos exists only as long as those information pathways exist. It is a living system, which means that it does not exist as *a priori* substance, only as a dynamic ecology. Neither information pathways nor cyberethos is independent of or prior to the other. Cyberethos and pathways are mutually constitutive, both cause and effect at the same time.

SUGAR-CUBE HEADS
THE MATERIAL LIMITS ON CYBERETHOS

The existence of cyberethos as a "living system" highlights the material as well as the discursive forces at play in the creation of good character, pointing us to the importance of bodies, places, and communities, both local and global, in the constitution of good character. The role of discourse, of coding in its various permutations and combinations, dominates accounts of cybernetic realities and subjectivities, as well as Aristotle's description of *ethos*. It would be easy to reduce cyberethos (or any system) to a discursive or a textual construct: all life becomes lines of code; all cyberethoi are products of speech as it is spoken. Ostensibly, Aristotle has something like this in mind with his comment that *ethos* "should result from the speech, not from a previous opinion that the speaker is a certain kind of person" (1.2.4). Cyberethos, then, would be meaningful only as a discursive pattern. Potentially, anything that can be said can be perceived as virtuous. Such positioning resonates to Weiner's musings "that it is conceptually possible for humans to be sent over a telegraph wire" because the only essential (human) attributes are those that can be coded (and transmitted) as information (*God* 36). However, reading Aristotelian *ethos* through Bateson's cybernetic lens reveals that good character is comprised of more than the single level. The flow of information out of which good character constitutes itself is subject to the materiality of its performance and its medium, constraints that inform and are informed by similar material constraints in Batesonian cybernetics.[5]

As Bateson points out, cybernetics is the science of constraints (*Steps* 399). It does not operate and explain by means of causal relationships. It operates and explains by means of specifying constraints. The materiality of cybernetic information is one such constraint, tying us to material sites and media, creating an identity that exists within the uncertain interface between discursivity and materiality, neither one nor the other, but both. The "difference that makes a difference" is a double-sided, double-faced process in which the ordering of language and the flux of reality constrain each other. In an elaboration of her father's work, Mary Catherine Bateson explains that physical reality—blood, bone, and earth—can be separated from the constitution of difference only at the level of description. All difference exists within and through the physical world: "We can meet the two [materiality and difference] only in combination, never separately" (Bateson and Bateson 18). Difference requires arrangements of matter, areas where materiality is "characterized by organization which permits it to be affected by information [news of difference] as well as by physical events" (18). On the other hand, physical reality requires the ordering of information. To be "known," however partially, however constructedly, reality must be represented. It must be coded, which by its nature affects that which is coded. Code and reality are fused. As Carol Gigliotti argues, even digital existence, the ultimate coded existence, is about bodies as well as codes; therefore, bodies must be central to our discussions of cybernetic subjectivities and ethicality:

> At the heart of all ethical and aesthetic investigations is the fact of our embodiment. . . . The point about the current insistence on the centrality of the body in discussions concerning aesthetics and ethics is not that we must transcend this envelope of the skin in order to act morally, but that the body allows us to be of and in the world at the same time. (55)

N. Katherine Hayles concurs. A hacker might dream of a life and of information free from material constraints, from what Case, William Gibson's protagonist in *Neuromancer*, so disparagingly refers to as the "meat." But "for information to exist, it must *always* be instantiated in a medium [. . .]. [A]bstracting information from a material base is an imaginary act" (75).

Considering Aristotelian *ethos* through a cybernetic lens directs our attention to a similar reciprocity between materiality and discourse. According to Aristotle, *ethos* is comprised not only of *phronesis* and *arete*, but also of *eunoia*, an element of *pathos*, or emotions, which ties *ethos* to the constraints of the material, to life-lived, not just life-spoken. *Eunoia*, or the feeling of good will between rhetor and audience that Aristotle claims is an element of *pathos*, is achieved by means of such material concerns as neatness, conviviality, and sympathy (see 2.4.11–29). The borders between the *pisteis* of *ethos* and *pathos* are not discrete; they are permeable. Thus, bodies, the locus of *pathos*, impinge on good character. Any consideration of the kinds of character an audience might possess requires that the rhetor contemplate what the auditors are like "in terms of emotions and habits and ages of life and fortune [*tyche*]," Kennedy says in his introduction to Book 2. *Eunoia* is firmly lodged in life-lived: "A friend is one who loves and is loved in return, and people will think they are friends when they think this relationship exists mutually" (2.4.2). An audience responds with friendliness to a speaker who is pleasant, good tempered, sympathetic, and appropriately attired. Thus, *ethos* is not discourse divorced from the parameters of the material. It is language and life combined, constrained by both. As such, neither identity nor character stops at a rhetor's skin or a city state's gates. It extends to include all the pathways communicating information, thereby constituting *ethos* as a system—inside and outside arbitrary boundaries of flesh, medium, and environment.

Cyberethos, then, is a living system materially constrained, and as such it offers insight into ethical dilemmas resulting from the permeable borders between words and flesh, especially within digital environments. The 1992 rape in cyberspace and the ensuing discussion over the ethicality of that rape offers an apt illustration of a bread-and-butter-fly experience that can be beneficially analyzed from the perspective of a cyberethos. I return to an incident more than a decade old to underscore the degree to which the questions raised in that situation continue to plague our cybernetic realities. While the digital technologies—or the interfaces mediated between technologies and users—may have radically changed in the last decade, the dilemmas that plagued LambdaMOO in 1992 align with dilemmas my fifteen-year-old daughter confronts on Gaia, established in 2003, as she engages in creating and responding to narrative actions written by participants linked only by an interactive website. While Anna has the power to build and insert a character into a storyline, she cannot control what other participants might do to or with her character. Cyberethos provides a robust framework for critically analyzing online choices and the justifications for those choices.

In 1992, in LambdaMOO, one of the oldest existing MOOs, an avatar named Mr. Bungle used a voodoo doll, a phantom program that co-opts a fellow participant's character by overwriting that character's discourse, to control the actions of two other avatars in the MOO. Mr. Bungle forced these enslaved characters to perform a variety of sexual acts on him, themselves, and one another. Even after he was ejected from the room, Mr. Bungle continued the assaults until he was finally barred ("toaded") from the system and denied reentry, a kind of virtual death sentence (Dibbell 239–40, 245; Turkle 249–54). One of the victims later responded to the assault by demanding Mr. Bungle's "virtual castration" while simultaneously complaining about his lack of civility (qtd. in Dibbell 243).

During an ensuing online discussion of virtual rape, one self-styled MUD rapist defended the activity. Virtual rape, he says, is something done "in a free non-meaningful manner," something that although it is "plain out sick" is done to have "fun" (qtd. in Turkle 253). VR "hurts" no one; therefore, rape should not elicit censure. However, throughout the incident and the follow up dialogue, VR and RW tangle in inextricable ways, from the equation of lost civility with virtual rape to the confusion of victims. As the self-styled MUD rapist reveals, "There are other MUDs where we have done the same thing and even though the victim didn't like it, the GODs told the victim 'too bad, it's not like they pkilled [player killed] you'" (qtd. in Turkle 253). Obviously, in contradiction to his previous comment in the same posting, the discussant is aware that VR rape does hurt the victim, but it is an offense less heinous than pkilling the victim.

Within this discourse and within these actions, where do we locate identity or good character? Do we focus on the actions and discourse in the virtual sphere? Or do we turn our attention to the individual, embodied user who, willy nilly, brings to that virtual sphere the cultural predispositions, assumptions, and prejudices that hold sway in the RW?[6] From a cyberethos perspective, the answer is no to each of these questions. We cannot locate identity or good character within the avatar because the avatar exists only by means of a user's graphical-discursive transactions within a particular environment. Nor can we locate identity within the embodied user—similar to a rhetor's life outside of the discourse—because we do not know the embodied user except through the avatar's discourse. Rather, identity consists of a mutually constitutive cyberethos materially constrained. Assigning identity to or determining good character based solely on the avatar, the user, or the discourse compromises the integrity of the living system. Cyberethos points us, instead, to the entire ecological context as the site for identity and good character.

Cyberethos is sited on the cusp between the life-lived and the life-spoken. It is only when we understand good character as both rhetorical and material that we can understand why one victim mourns both the rape of the discourse, through the loss of civility, as well as the rape of the avatar, through that uncivil discourse.

It is only when we see good character at the level of a rhetorical-material nexus that we can understand why the individual defending rape on a listserv would offer a postscript apology for his grammatical errors and formatting problems, issues of style and delivery that the user implicitly feels undercut the persuasive appeal of his discourse. RW and VR twist into a mobius strip where one side cannot be distinguished from the other. As journalist Julian Dibbell notes, the complaints about Mr. Bungle's incivility as well as calls for his dismemberment, "[l]udicrously excessive by RL's [real life] lights, woefully understated by VR's" make "sense only in the buzzing, dissonant gap" between RW and VR, or in the buzzing, dissonant gaps within which the rhetorical and material levels of a cyberethos organize themselves (243). Cyberethos requires us to attend to the material elements of any construction of identity, of good character. From the perspective of cyberethos, any identity that ignores or erodes sensitivity to material constraints is unethical.

"IT HAPPENS ALWAYS"
THE IMPACT OF TIME ON *ETHOS*

The material constraints on cyberethos highlight the performative aspect of any system of good character: identity exists within a span of time as a span of time (and it depends on the medium of the moment). Reading Aristotelian *ethos* through the lens of cybernetic time offers insight not only into the performative aspect of cyberethos, but also into the continued existence of cyberethos across performances, balancing evanescence with permanence, improvisation with habit. Time impinges on all aspects of the living system; it constrains the evolution, survival, and death of identity, highlighting a similar dynamic in bread-and-butter-fly realities that evolve and dissolve.

Cybernetic time is marked by its nonlinearity, a characteristic that directly impinges on the constitution of good character. Cybernetic systems are premised on nonlinear temporal patterns and, thus, are irreversible in time. Dating back to the sixteenth century, time has been perceived as linear and sequential, a perception strengthened by the evolution of the clock that parsed experiential time into discrete units. Undergirding linear time are two linked, although counterintuitive, concepts: cause and effect proceed in an orderly fashion; and time, because of this orderly progression, is reversible. Newtonian physics hypothesizes that every event is caused by some specific initial condition of particles in the universe. That initial condition causes an equal effect that then becomes a cause of another equal effect, proceeding in a retraceable chain from cause to effect. If we can determine the precise initial conditions of an event, then we can predict the exact outcome by tracing the intertwining strands of cause and effect in the rope of time. Theoretically, if all initial conditions—the position and momentum of all particles—can be ascertained, then we can discover everything that will occur or that did occur, back to the first cause of God as the prime mover. From this orientation, neither past nor present nor future is relevant because we are living in a present moment that encompasses past (initial conditions) and future (determined outcome). In a universe governed by immutable laws of motion and of linear cause-effect relationships, time is essentially reversible.

From the perspective of linear time, if we identify the initial cause or the starting point for good character, then we can predict and control its persuasive effect. However, cybernetic systems do not function according to linear time. Rather, both are probabilistic. In *The Human Use of Human Beings*, Wiener writes that the groundwork for a probabilistic theory of messages rests with the discovery that initial causes cannot be determined with anything except probability. Uncertainty is an inextricable element of our physical and intellectual existence. Therefore, without the ability to ascertain the precise nature of an initial event, we cannot predict ensuing events with any accuracy. The process by which a user accesses a website (or any hypertext) and its various links illustrates the difficulty of locating an initial cause. A user need not approach a website from its homepage, its designated starting point, and then proceed in an orderly fashion from link to link. Instead, a user can enter that site through the "back door" (or the "side door") by connecting to it from a second website or directly accessing a link site, entirely frustrating an effort to control a user's position and momentum. Nor can the website designer predict or direct the way in which a user might meander through the links of a site (or whether the user clicks in and out of that site). Unable to determine a starting point, the designer cannot extrapolate the nature of the *ethos* a user creates of the site and of the site designer. This process is rendered even more complex when we factor in the nonlinearity of cause and effect in cybernetic systems.

Instead of a small cause resulting in a small effect, small causes in cybernetics can result in very large effects. A butterfly flapping its wings in Buenos Aries can contribute to a hurricane off the coast of Florida (Gleick). The change in a single digit of code (or the bumbling flight—or desiccated corpse—of a moth) can crash programs. The appearance and disappearance of links and websites reconfigure the Internet. Capturing the volatility of cybernetic time and cybernetic identity, Peter Lunenfeld calls this the digital "state of unfinish": "The digital dérive [drifting] is ever in a state of unfinish because there are always more links to create, more sites springing up every day, and even that which has been catalogued will be redesigned by the time you return to it" (10). The dissolution and resolution of authorial *ethos* in cybernarratives highlight the way in which minimum causes can have maximum effects in cyberspace's state of unfinish.

In the fifteenth century, the rise of moveable type and the increasing stability of the printed text marked a parallel rise and increasing stability of the author position. At one point in both critical history and in the evolution of individual creative works, to know the poem, one had to know the author because, as fifteenth century European copyright laws suggested, the author "owned" the poem. Now, the good character of owner and ownership are continually short circuited, theoretically by the inroads of postmodernist stances and concretely by the experience of cybernarratives, which deliberately court nonlinear cause and effect patterns. For example, via online, interactive sites such as Gaia or through software such as Storyspace—a hypertextual narrative program that invites readers to intervene by contributing to, subtracting from, or rearranging the text—a small contribution by a single user holds the potential to reconfigure the authorial good character of the entire site. Small inputs result in maximum outputs.

Hypertext novelist Carolyn Guyer describes this phenomenon. She notes her pained recoil when a writer, one she guiltily deems less proficient than she, enlarges her hypertext novel at her invitation in ways Guyer finds unsatisfactory. In spite of her desire to have readers add to her work, Guyer is unprepared for the extent to which one small addition to a much larger array of lexia (text blocks) results in the reformation of the good character of the entire site. Her visceral rejection functions on two levels: first, she rejects the specific textual emendation, and, second, she recoils from the massive reconfiguration of the entire work effected by that specific textual addition.

Reading Aristotelian *ethos* through the lens of cybernetic times offers us a framework for understanding the volatility as well as the permanence of good character. Conceived cybernetically, Aristotle's *ethos* can be interpreted as nonlinear, lodged as it is within the realm of probabilistic or contingent knowledge. Rhetorical time is in a similar state of unfinish, and Aristotelian *ethos* is subject to the same difficulty in determining its initial conditions. James Baumlin and Tita French Baumlin highlight this problem. They describe the evocation of *ethos* as a series of feedback/feedforward relationships between rhetor and audience that are analogous to the circularity of cybernetic time. Drawing on Freudian psychology, they argue that *ethos* is a projection of a listener/reader's superego. The listener/reader brings an array of predispositions, biases, expectations, and orientations to the speech act. It is the baggage she carries. The rhetor then offers cues that transact with these predispositions to summon forth a particular shape from the listener/reader's expectations, which the audience member validates by sampling the rhetor's cues. Within this temporal circularity, we cannot determine initial conditions; therefore, we cannot predict the specific evocation that *ethos* will take, if it takes any. *Ethos* depends on an audience's predispositions, as Baumlin and Baumlin point out, but those predispositions exist as potential, not as actual, systems. The specific predispositions contributing to the evocation of *ethos* do not exist until that *ethos* is evoked, harkening back to the constitution of *ethos* as a living system. Predispositions come to being within the evocation of the entire array of information pathways. The concept of initial conditions—as something existing outside of the system—is foreign and inapplicable to *ethos*. We cannot point to a time prior to the evocation of *ethos*, or to an event that exists "before" *ethos*, because neither time nor event is relevant outside of network of relationships constituting *ethos*. Instead, *ethos* comes to be within an instant as an instance (Metzger).[7]

In addition to the impossibility of ascertaining the initial conditions of *ethos*, overlapping time in rhetoric allows an effect to be its own cause. A circular process is at play whenever communication occurs. When a rhetor offers something intended as serious, it is because she perceives herself as serious. However, if the audience interprets the cue as amusing and laughs, the rhetor sees that response as a rendering of herself as amusing. She accepts that judgment and incorporates such humor as part of her system of *ethos*, a process that creates of the audience an ethical construct. She creates her system of *ethos* because of the audience's *ethos*, offering cues that validate both the audience's construction of her newly realized *ethos* and her construction of the audience's good character. All of this takes time for news of *ethos* to travel throughout the pathways. In fact, a system of *ethos* evolves out of the lapse of time, allowing for and necessitating the minute adjustments that contribute to the maintenance of the entire performance. Thus, because time loops back on itself, a system of *ethos* can paradoxically give birth to itself.

This cybernetic reading of Aristotle's *ethos* returns us to the intricacies of good character and enables us to reframe ethical authority. For example, the interactivity of cyberethos is an integral aspect of cyberspace narrative. Thus, the concept of a master text or a master writer loses its good character because there is no master text or master writer. Readers become writers as easily as they become readers—wreaders—and out of the ashes of one evolves the substance of the other. The location and identity of wreader in Storyspace, for instance, shifts with time (and with space) depending on the lexia (text block) viewed, the time viewed, and the particular user interacting. On Gaia, identity shifts from story thread to story thread, from forum to forum.

Furthermore, its own self-organization changes the grounds and conditions of a wreader's existence. The evocation of a cyberethos becomes information that then feeds back into the system, simultaneously constituting it and setting out the terms of its own degradation. Creating an interactive hypertext novel evokes a particular authorial identity, one that Barbara Page describes as a feminist literary identity and containing the seeds of its own dissolution. When a wreader responds to an invitation to write to an author's work, the initiating authorial identity no longer exists, just as the locus of ownership and the stability of the text are rendered problematic. The process of constructing an identity based on the openness of interactivity means that the act of interacting will change the nature of identity. The performance of such a cyberethos requires its own disappearance: the sugar cube head dissolves in the weak tea.

Finally, the dissolution of one identity yields a space for the constitution of another. For instance, within cyberspace, authorial identity and master text yield to the good character of the "wreader" and the "restive text." Like the slippery temporal separation between Aristotle's audience and rhetor, the temporal nonlinearity so characteristic of cyberspace interactivity destabilizes the boundaries separating reader and writer, creating the authority of the wreader out of these shifting boundaries. Similarly, Page argues that hypertext's state of unfinish manifests the restive text, one that refuses a point of reference stable across either time or space. The dissolution of identity at the hypertextual site resolves itself into a cyberethos whose good character is predicated on that very dissolution, a good character that challenges the traditional authority of patriarchal structuring.

To slightly redirect Johnson-Eilola, such an approach moves identity and ethics "out of simple cause and effect toward an understanding of culture and technology as contingent, multidimensional, fragmented, and constructed in local uses rather than universal determinations" (9).

"THEN IT WOULD DIE, OF COURSE" CYBERETHOS AND ETHICS

As a living system, constrained by both materiality and nonlinear time, cyberethos provides a framework for understanding identity and good character in a bread-and-butter-fly age. But it provides more than a framework, important though that might be; cyberethos also offers us guidelines for ethical action. Cyberethos spins out of the significant differences constituting its own system, leading us to a definition of ethical judgment that relies on three strictures: the need to discern the fluid, dynamic, and constructed nature of a cyberethos in which we are producer and produced; the need to ensure the continued fluidity of those boundaries by taking responsibility for them; and the need to recognize the material and temporal constraints imposed on and enacted by any cyberethos. From the stance of cyberethos, we are called to act on and judge our inescapable deterritorialization, materiality, and temporality, all of which necessarily deploy good behavior as well as good character throughout the constantly shifting pathways of a system. Like cyberethos, we cannot reduce ethics to language severed from life or life severed from language, divorcing both from place and time. We cannot return to an autonomous subject, to a reader separated from a writer. Instead, we must account for the interactivity and destabilization of our lives and identities, finding in that interactivity and destabilization our ethical tenets. Cyberethos draws us to a cybernetic aesthetic, where beauty and goodness are implicit within the act itself, not in some intended effect (see Bateson, *Sacred* 253-57). As Aristotle explains, "we call complete without qualification that which is always desirable in itself and never for the sake of something else " (NE 1097a30).

Whereas cyberethos establishes a good character, a persuasive way of facing the world, ethics acts on that good character, a double process of doing and judging. To the Aristotle of *Nicomachean Ethics*, ethics is a performance, and that performance slips between judging and doing. It is a judgment in and of the doing, judgment as doing. Because ethics exist in the activity, the pleasure of a virtuous life is in the performance of virtuous acts (NE 1099a 10), not in the understanding of virtuous acts (see Engberg-Pedersen 121). Capturing this transactivity of ethics, Aristotle explains that possessing virtue or understanding of virtue is inseparable from acting virtuously: "for one who has the activity will of necessity be acting, and acting well" (NE 1099a 3-5). One cannot believe without also acting. Although ethics cannot be equated to cyberethos, neither can it be separated from it because they are part of the same living system. Cyberethos yields an ethical depth perception that enables us to assess virtuous behavior as we enact both behavior and judgment. It is an analytical stance that cannot be abstracted from the system it analyzes, for the judge is always a part of that which is judged as well as a part of the judgment. Good character and good action are situated in actively shifting pathways, the behavior of which yields an ethics of performance constantly morphing between doing and judging.

There is a pleasure in the freedom of constantly shifting, permeable boundaries, Haraway points out, and we should revel in that freedom of systemic self-creation, self-generation, self-adaptation. But simultaneously that pleasure requires that we take responsibility for the boundaries that we inevitably make and strive to maintain (*Simians* 150). With an acknowledgment of cyberethos as a living system that functions within the constraints of time and place, we can better understand the ways in which a Napoleon or a Hussein can be part of a system within which they and their actions are viewed as wise, virtuous, and of good will. Also, we can better understand our complicity in those systems. Through considerations of the flow of differences that make a difference, we can ask how a system becomes pathological, creating a cyberethos that becomes toxic to its own existence as a system, as in terrorism, racism, or genocide. On one level, a cyberethos generates its own identity, strives after its own self-continuation through the constant do-si-do of its constituent elements. What serves that balance, serves the system, and thus is virtuous. What is pathological—or unethical—is that which threatens the good character of that portion of a system. However, systems do not exist in isolation. They are tied to, constrained by, larger and smaller loops. What is good for cyberethos at one level can become toxic to the system at another. Thus, terrorism requires adaptive measures that will in effect destroy the larger system within which it is immanent. Like bread-and-butter-flies, its end is consumed by its means.

In addition to understanding the toxicity of a single loop in a larger system that is pathological to the whole, the ethical perspective we gain through cyberethos also enables us to understand instances where good character requires the death of the very system that gives it its existence. For example, in his 1974 speech for the Governor's Prayer Breakfast, Bateson tells the story of a group of Native Americans who were under siege to discontinue the use of peyote in their religious ceremonies. A leading anthropologist proposed to the tribe that he film their ceremonies as a means to demonstrate the religious nature of the drug use. However, to allow the filming, which may have led to the continued legal use of peyote, the Native Americans believed that they would be destroying the integrity of the ceremony they wished to preserve. They refused to allow the filming. "In the story," Bateson says, "the Indians perceive that it is nonsense to sacrifice integrity in order to save a religion whose only validity—whose point and purpose—is the cultivation of integrity. The Indians declined to save their religion on those terms" (Bateson and Bateson 75).

Cyberethos and the ethical stance that emerges from it offer us insight into and a framework for acting in the bread-and-butter-fly age. Cyberethos offers a means to unite ourselves in an ecology of good character, reconfirming/recreating our identity and connectedness within the larger systems that offer us life, and assuming responsibility for that identity and that life. Nothing is more monstrous, Bateson says, than the attempt to separate the mind from the body, the external mind with the internal mind: "When you separate mind [or cyberethos] from the structure in which it is immanent, such as human relationships, the human society, or the ecosystem, you thereby embark, I believe, on fundamental error, which in the end will surely hurt you" (*Steps* 470, 393). Marked by permeable boundaries, materiality, and nonlinear time, cyberethos demands that we link our survival (as a species and as rhetors) with our good character. There can be no survival without good character. To slightly misquote Bateson, "Arrogate all mind [cyberethos] to yourself, you will see the world around you as mindless and therefore not entitled to moral or ethical consideration" (*Steps* 468). The environment—the discourse situation—becomes ours to exploit. If the survival unit is interpreted as us and our loved ones against a hostile world, Bateson says, we will destroy the world and ourselves with our own hate. That which feeds us also consumes us. But cyberethos offers us a way out of the us/them dichotomy because at the moment of its evocation us and them are indivisible. Perhaps virtue, good will, and practical wisdom translate into communal and individual survival, which is, after all, the ultimate means of persuasion.

NOTES

1. The crossover between Aristotle and twentieth-century cybernetics has been explored by various scholars. See, for example, Lawrence William Rosenfield's examination of Aristotelian and cybernetic causality.

2. My project resembles in many aspects James Porter's agenda in *Rhetorical Ethics and Internetworked Writing*, where he proposes an ethical framework—his rhetorical ethics—for addressing dilemmas raised by the phenomenon of internetworked writing. Porter's rhetorical ethics, drawn from rhetoric and postmodern ethics, provides "guidance in the form of general principles . . . as well as in the form of procedural strategies" (xiii). However, I am concerned in this essay with cybernetic realities that extend beyond internetwork. Also, I arrive at my framework through a fusion of classical rhetoric, specifically Aristotle's concept of *ethos*, and Bateson's ecological cybernetics. The degree to which Bateson's work aligns with postmodern thinking (see Lévy; Deleuze and Guattari) is beyond the scope of my argument.

3. Focusing exclusively on new information technologies, Barbara Warnick claims that neoclassical rhetoric, especially Aristotelian rhetoric, is unsuitable for exploring the challenges of this cybernetic era. She advocates, instead, the use of new critical theory. In contrast, other rhetorical critics have emphasized the usefulness of neoclassical rhetoric for theorizing a reality of blurred boundaries, turning to sophistic rhetoric (Ballif) or Isocrates (Welch). What these critics hold

in common with Warnick is their dismissal of Aristotelian theory for understanding our current cybernetic reality. However, as I illustrate here, Aristotle, read through a cybernetic lens, offers unique insights into the nature and challenges of a bread-and-butter-fly reality.

4. See Guilbaud for the derivation and history of the term *cybernetics*.

5. The inextricability of form and content in Aristotle is illustrated practically in his *Poetics* and philosophically in *De Anima* through his concept of hylomorphism.

6. See Kang on the presence of racial prejudice within MOOs, as well as on the possibilities of cyberspace for furthering racial justice.

7. Time is an integral part of rhetoric. James Murphy defines rhetoric as "the study of the means for producing future discourse" (75); Aristotle discriminates between three kinds of rhetoric—epideictic, judicial, and deliberative—on the basis of time, associating each with a particular time: epideictic with the present moment, judicial with a past action, deliberative with a future action (1.3.4).

WORKS CITED

Aristotle. *De Anima*. Trans. J.A. Smith. *The Complete Works of Aristotle: The Revised Oxford Translation*. Ed. Jonathan Barnes. Vol. 1. Bollingen Series 71.2. Princeton: Princeton UP, 1984.

———. *Ethica Nicomachea (Nicomachean Ethics)*. Trans. W.D. Ross. *Introduction to Aristotle*. 2 ed. Ed. Richard McKeon. Chicago: U of Chicago P, 1974. 338–581.

———. *On Rhetoric: A Theory of Civic Discourse*. Trans. George A. Kennedy. New York: Oxford UP, 1991.

———. *Poetics*. Trans. I. Bywater. *The Complete Works of Aristotle: The Revised Oxford Translation*. Vol. 2. Ed. Jonathan Barnes. Princeton: Princeton UP, 1985.

Ballif, Michelle. "Writing the Third-Sophistic Cyborg: Periphrasis on an [In]Tense Rhetoric." *Rhetoric Society Quarterly* 28 (1998): 51–72.

Bateson, Gregory. *Sacred Unity: Further Steps to an Ecology of the Mind*. Ed. Rodney E. Donaldson. New York: HarperCollins, 1991.

———. *Steps to an Ecology of the Mind: Collected Essays in Anthropology, Psychiatry, Evolution, and Epistemology*. 1972. Northvale, NJ: Jason Aronson, 1987.

Bateson, Gregory, and Mary Catherine Bateson. *Angels Fear: Towards an Epistemology of the Sacred*. New York: Macmillan, 1987.

Baumlin, James S. "Introduction: Positioning Ethos in Historical and Contemporary Theory." *Ethos: New Essays in Rhetorical and Critical Theory.* Ed. James S. Baumlin and Tita French Baumlin. Dallas: Southern Methodist UP, 1994. xi–xxxi.

Baumlin, James S., and Tita French Baumlin. "Psyche/Logos: Mapping the Terrains of Mind and Rhetoric." *College English* 51 (1989): 245–61.

Bolles, Edmund Blair. *A Second Way of Knowing: The Riddle of Human Perception.* New York: Prentice, 1991.

Carroll, Lewis. *Through the Looking-Glass and What Alice Found There.* 1871. New York: Quality Paperback, 1994.

Deleuze, Gilles, and Felix Guattari. *A Thousand Plateaus: Capitalism and Schizophrenia.* Trans. Brian Massumi. Minneapolis: U of Minnesota P, 1987.

Engberg-Pedersen, Troels. "Is There an Ethical Dimension to Aristotelian Rhetoric?" *Essays on Aristotle's Rhetoric.* Ed. Amélie Oksenberg Rorty. Berkeley: U of California P, 1996. 116–42.

Gibson, William. *Neuromancer.* New York: Ace, 1984.

Gigliotti, Carol. "The Ethical Life of the Digital Aesthetic." *The Digital Dialectic: New Essays on New Media.* Ed. Peter Lunenfeld. Cambridge, MA: MIT P, 1999. 46–63.

Gleick, James. *Chaos: Making a New Science.* New York: Penguin, 1987.

Gromala, Diana J. "Virtual Avatars: Subjectivity in Virtual Environments." Visible Language 31.2 (1997): 214–29.

Guilbaud, G.T. *What is Cybernetics?* Trans. Valerie MacKay. New York: Criterion, 1959.

Guyer, Carolyn. "Fretwork: ReForming Me." http://mothermillennia.org/Carolyn/FretworkI.html

Haraway, Donna, J. *Modest_Witness@Second_Millenium.FemaleMan©_Meets_OncoMouse™: Feminism and Technoscience.* New York: Routledge, 1997.

———. *Simians, Cyborgs, and Women: The Reinvention of Nature.* New York: Routledge, 1991.

Hayles, N. Katherine. "The Condition of Virtuality." *The Digital Dialectic: New Essays on New Media.* Ed. Peter Lunenfeld. Cambridge: MIT P, 1999. 69–94.

Johnson-Eilola, Johndan. *Datacloud: Toward a New Theory of Online Work.* Cresskill, NJ: Hampton P, 2005.

Kang, Jerry. "Cyber-Race." *Harvard Law Review* 113 (2000): 1130–208.

Lévy, Pierre. *Becoming Virtual: Reality in the Digital Age.* Trans. Robert Bononno. New York: Plenum, 1998.

Lunenfeld, Peter. "Screen Grabs: The Digital Dialectic and New Media Theory." *The DigitalDialectic: New Essays on New Media.* Ed. Peter Lunenfeld. Cambridge: MIT P, 1999. xiv–xxi.

McDonough, Jerome P. "Designer Selves: Construction of Technologically Mediated Identity within Graphical, Multiuser Virtual Environments." *Journal of the American Society for Information Science* 50 (1999): 855–69.

Metzger, David. *The Lost Cause of Rhetoric: The Relation of Rhetoric and Geometry in Aristotle and Lacan.* Carbondale: Southern Illinois UP, 1995.

Murphy, James J. "What Is Rhetoric and What Can It Do For Writers and Readers?" *The Reading-Writing Connection: Ninety-seventh Yearbook of the National Society for the Study of Education.* Part 2. Ed. Nancy Nelson and Robert C. Calfee. Chicago: U of Chicago P, 1998. 74–87.

Page, Barbara. "Women Writers and the Restive Text: Feminism, Experimental Writing, and Hypertext." *Cyberspace Textuality: Computer Technology and Literary Theory.* Ed. Marie-Laure Ryan. Bloomington: Indiana UP, 1999. 111–36.

Porter, James E. *Rhetorical Ethics and Internetworked Writing.* Greenwich, CT: Ablex, 1998.

Reeve, C.D.C. "Philosophy, Politics, and Rhetoric in Aristotle." *Essays on Aristotle's Rhetoric.* Ed. Amélie Oksenberg Rorty. Berkeley: U of California P, 1996. 191–205.

Rosenfield, Lawrence William. *Aristotle and Information Theory: A Comparison of the Influence of Causal Assumptions on Two Theories of Communication.* The Hague: Mouton, 1971.

Ryan, Marie-Laure. "Introduction." *Cyberspace Textuality: Computer Technology and Literary Theory.* Ed. Marie-Laure Ryan. Bloomington: Indiana UP, 1999. 1–28.

Smith, Alvy Ray. "Digital Humans Wait in the Wings." *Scientific American* 283.5 (2000): 72–78.

Turkle, Sherry. *Life on the Screen: Identity in the Age of the Internet.* New York: Simon, 1995.

Warnick, Barbara. "Rhetorical Criticism in New Media Environments." *Rhetoric Review* 20 (2001): 60–65.

Welch, Kathleen E. *Electric Rhetoric: Classical Rhetoric, Oralism, and a New Literacy.* Cambridge: 1999.

Wiener, Norbert. *The Human Use of Human Beings: Cybernetics and Society.* New York: Da Capo, 1954.

——. *God and Golem, Inc.: A Comment on Certain Points where Cybernetics Impinges on Religion.* Cambridge: MIT P, 1964.

2
SCREENING (IN)FORMATION

BODIES AND WRITING
IN NETWORK CULTURE

Jennifer L. Bay

In *The Language of New Media*, Lev Manovich declares that we live in "a society of the screen" (94). While the screen has been used to frame visual representations for centuries, Manovich explains, "Today, coupled with the computer, the screen is rapidly becoming the main means of accessing any kind of information, be it still images, moving images, or text" (94). Human beings in the twenty-first century are constantly surrounded by screens and screening devices. Security cameras record and screen our movements in almost every public place or business. At our workplaces, we are constantly seated in front of computer terminals during the day, staring at screens that display text and image. At home, we encounter another framed screen that projects television shows, movies, and video games. We interface with video display terminals at ATMs, information kiosks, public telephones, grocery store checkouts, gas stations, and more. In almost every environment, we interact with some kind of screening device that mediates our experiences as human beings who read, write, and think. Considering this pervasive environment, it is curious that we so seldom speak of these screens in our work on rhetoric.[1]

Against this background of screens emerges Mark C. Taylor's text, *The Moment of Complexity: Emerging Network Culture*, which provides us with one glimpse of how screening is enmeshed with writing and rhetoric. For Taylor, writing always involves screening information, but Taylor's screen is both a net that filters and a surface for projection. This "both/and" description of screening is essential to Taylor's model of complexity theory since his concern is with the ways that everything and everyone is inextricably enmeshed in a constantly changing network, which is made possible by information processes. Taylor theorizes that we live in a world where new technologies are changing the way that we perceive information: "Information is not limited to data transmitted on wireless and fiber-optic networks or broadcast on media networks. Many physical, chemical, and biological processes are also information processes" (4). In order to account for these new information processes, Taylor's book attempts to theorize "the relation between nature and culture in such a way that neither is reduced to the other but that both emerge and coevolve in intricate relations" (4). Taylor's phrase for that relation is *complex adaptive networks*. Such networks—or network logics—are embodied in human beings. As Taylor claims, "We are gradually discovering that we are, in effect, *incarnations* of worldwide webs and global networks whose complexity is fraught with danger as well as opportunity" (17). Thus, the screening that happens when we write comprises who we are as embodied subjects.

Taylor's model is clearly useful for rhetoricians in thinking about communication in network culture. However, less apparent is the connection to Manovich's observations about our relationships with the physical screens that pervade our lives. Manovich asks us, "What are the relationships between the physical space where the viewer is located, her body, and the screen space?" (94). Substitute *viewer* with *writer* in this quotation, and the question becomes one of rhetorical situation. While Taylor's model of complexity theory helps us to understand the importance of information processing, the material conditions—the human bodies—that put information into motion remain largely invisible. In light of Manovich's assertions, how do our bodily interactions with physical screens function as rhetorical components in Taylor's model of complexity?

In order to answer this question, we need to look at more than just *The Moment of Complexity*. One of Taylor's earlier works specifically deals with the body and can help us to think about the relationship of the body to the screen. That book is *Hiding*. Drawing from research in Internet studies, body studies, performance art, and critical theory, I will explore the implications of a rhetoric that, in Taylor's language, "simultaneously embodies and articulates the *incarnational logic of networking*" (230). Moving from local examples of bodies as individual networks to more global examples of bodies networking, I will argue that embedded in Taylor's texts is a notion of rhetoric that can provide us with exciting new visions for embodied rhetorics.[2] Such a vision would be grounded in the notion that the body and computer together form a complex, co-adaptive system. At the end of this essay, I will suggest that recent work on affect, such as by Brian Massumi, supplements Taylor's model by attending to movement, but that movement does not take into account the technology behind movement. It is not enough to posit a theory of bodies, movement, and affect without confronting our bodily connections with machinery on a daily basis.[3] Thus, for instance, while the concept of the cyborg is useful in thinking about bodies and/as machines, it needs to be extended to account for the co-adaptive, systemic relationships that humans have developed with information processing and viewing machines. If we are to think about bodies in a meaningful way, we must examine the body's relationship to the computer screen and the body's screening of online information.

Taylor's complexity model provides theoretical grounding for the material ways that bodies both become screens and interact with a variety of screens on a daily basis. The human body screens information that it receives through sensory apparatuses, bodily filters, neural activities, and cognitive operations, and yet it also functions as a screen on which culture and identity are projected; in this way, the body simultaneously functions as filter and surface. Moreover, our screening bodies are in constant contact with other technological screens, such as computers, television monitors, webcams, and digital cameras, prompting us to question the comportment of the human body in relation to these screens and screening devices. For instance, how are we to think about the body that uses the camera phone to take pictures of others *using a screen*? What then is the relationship between the body that uses the phone to talk, to write text messages, and to take pictures? If Iris Marion Young is correct in suggesting that there is a feminine body comportment, will there emerge a technological body comportment for those who have access to particular technologies?

As our bodies interact with these screens—as our bodies as screens confront other technological screens—the body becomes a reflected and refracted image; it is both filtered, filtering, and reflected back on itself. Taylor's work asks us to see this relationship between bodies and information as a complex adaptive system, to "see" ourselves as both information and screens. For instance, when we catch our reflection in the computer screen, we momentarily witness our body as it is screening information, and yet simultaneously it serves as another screen that processes the physical screen in front of us. The image that surfaces from this engagement seems to become another piece of information displayed on the screen. The material boundary between technological artifact and body is blurred, and the body simultaneously becomes image, information, and material object. This self-reflective screening forms a complex, co-adaptive system of inscription that transforms our view of the body's relationship to contemporary digital writing practices. While there have been recent calls for an "embodied rhetoric" in composition studies, those calls have often focused on bringing back or acknowledging the presence of the affective body in writing processes. However, viewing the body within a complex co-adaptive writing system requires rethinking the body as always in motion, always moving and shifting as a matrix of information processes. Thus, for example, the composition student's body does not merely "write" a paper using a computer; the student body and the computer form a complex co-adaptive writing system that produces both material affects and virtual effects. A techno-bodied subjectivity would consider the recursive processes of screening and reflecting that form our networked identities; we can most readily see those processes in online networking practices. But before we move to online networks, let's examine how Taylor prepares us to see bodies and information in his earlier book on the body, *Hiding*.

RHETORIC AS BODY SCREEN

Hiding, in Ellen Lupton's words, is "a commentary on the culture of skin" (31). A visual experience in itself, the physical book provides the reader with images of the body overlaid with sheer, skin-like pieces of paper. This material presentation, along with the theoretical text, allows Taylor to challenge the modernist binary of surface/depth and the corollary binary of appearance/essence. A commonplace of postmodern theory is that it challenges any stable relationship between these binaries, and for Taylor, this opens up creative potential for a world filled with surfaces. In Taylor's thought, the body possesses incredible new possibilities for inventive acts. Since the outer is no longer merely an expression of the inner, the body is no longer "a sign that can be deciphered by those who know the code" (15). When our theories rely on depth metaphors and models, we always seek what we cannot see, or what lies beneath; thus, the body becomes the cover, actualized in the skin, that must be stripped away in order to discover, to "see," the true self. But this model eliminates the possibility for seeing the body in productive ways, as well as for thinking about the body in an age of media saturation. As Taylor concludes, "By repeatedly seeking what hides, we tend to forget how to read the surfaces on and between which life is lived" (25). One of the primary surfaces that Taylor wants to remember is skin.

As a way to remember how to read surfaces, Taylor turns to transgressive and transformative skin practices that create and invent new possibilities for seeing the body outside of the surface/depth distinction. Tattooing, fashion, performance art, and new media technologies all interrogate and bring forth the skin as meaning (less). Taylor's discussion of these explicitly material events evokes the body in obvious ways. Skin functions as more than a barrier or surface; skin serves as a conduit, or canvas, for the performance of subjectivity and thought. What we see here in germinal form is the notion of body as screen, a notion born out in the concluding chapter of *Hiding*. There, Taylor explains, "The notion of *distributed intelligence* redraws the boundaries within the body, between mind and body, and between self and world. While the mind is not a mind, but a network of networks, the body is not a body, but, in Donna Haraway's terms, a 'network-body'" (323). By the time that *The Moment of Complexity* emerges, Taylor has reconceptualized the relationship between mind, body, and self in terms of complexity theory and adaptive systems.

One of the first places where we see this reconceptualization surface in *The Moment of Complexity* is in his discussion of writing. Consider how Taylor pushes us to understand the activity of writing as dynamic:

> The moment of writing is a moment of complexity in which multiple networks are cultured. If writing does not push limits to the tipping point, it is simply not worth the effort. The writing that matters disturbs more than it reassures; it drives authors as well as readers to the edge of chaos and abandons them. Writers

realize that the pleasure of the text is not the satisfaction it
provides but the dissatisfaction is engenders. (198)

Taylor positions writing within the intersection of multiple networks of influence that are not only textual, subjective, and cultural, but also biological, objective, and natural. In short, the influences can be both sides of the binary: natural and cultural. He explains that networks are what allow these distinctions between these binaries to become "porous screens" (199). Whereas in *Hiding* he is more concerned with the collapse of the binary, here Taylor is ready to acknowledge that the binary is not in fact a binary at all but part of a matrix structure formed using porous screens.

The concept of screen forms the key to Taylor's discussion, but as he observes, "*Screen*, which, of course, can be either a noun or a verb, is a strange word in which multiple meanings pass through each other without losing definition" (199). It can mean "to separate" or "to filter," as well as" to conceal" and "to protect." Thus, rather than see the screen as a barrier between opposites, Taylor theorizes the screen as "more like a permeable membrane than an impenetrable wall; it does not simply divide but also joins by simultaneously keeping out and letting through. As such, a screen is something like a mesh or net forming the site of passage through which elusive differences slip and slide by crossing and crisscrossing" (199–200). The screen functions as a dynamic node in a network of relations rather than as a barricade. This translates into an understanding of nature and culture not as binary opposites but as different points in a network of relations. Our whole understanding of the world, as predicated on opposites, is moving more toward a "strange loop" created by technologies of production and reproduction (72). Such technologies (like the screen) force us to look to alternate metaphors that move beyond binary opposites.

"But," Taylor continues, "a screen is also a surface on which images, words, and things can be displayed. Every surface is actually a screen that hides while showing and shows while hiding" (200). A simple example of this phenomenon is the computer display. While it projects the interface of the machine, it also hides the internal hardware that processes the information. Another example would be the holographic image, at one time a popular means of presenting images on postcards and other memorabilia. The holographic screen usually displays two different images depending on the angle at which one views the image. While it's easy to make the general observation that screens can perform two different functions simultaneously, Taylor seems more interested in what happens at the nexus of those two functions, or where the two functions meet. In Heideggerean fashion, Taylor hints at the possibility that the process of screening is the condition of possibility for being: "In network culture, *subjects are screens* and *knowing is screening*" (200). If screening is a condition for being and for subjectivity, then the intersection between the process of screening and the material surface that screens is a critical juncture for identity formation and for understanding networking. Taylor expands his theory to address identity formation: "The screening critical to channeling experience, articulating knowledge, and cultivating meaning occurs through *dynamic* patterns" (204).

The recursive nature of the screen allows us to consider every surface as a screen and every knowledge-building act as a process of screening. That process of screening, though, is not a pure act of interpretation and meaning-making; rather, there is always an effect to screening that is itself another screen:

> While a screening is a presentation of a film, video, CD-ROM, website, etc., screenings are remains, waste, and detritus. Screening is often a process of filtering, designed to purify by removing, excluding, or repressing what threatens to contaminate. What is screened, however, does not simply disappear but lingers as dangerous refuse, which is neither precisely present nor absent. (288 n.3)

This is what it means to live in a networked culture—bodies are modalities of image, information, and material object; bodies are formed through the complex adaptive processing of all three modalities simultaneously. Under this system, writing expands into a political, embodied movement that constantly implicates refracted and reflected bodies.

Taylor's model offers us the possibility that rhetoric is the screen, the permeable membrane on which meaning is made and through which meaning is filtered. Taking complexity theory into consideration means we can no longer envision rhetoric as merely verbal, visual, and oral. Rather, rhetoric is networked among all three components through the ultimate screening device that is the body. What this means is that perhaps our minds and brains are not only filtering information, but our bodies are as well. Our bodies are implicated in the rhetorical choices we make and the rhetorical capacities (to filter and serve as filters) that we possess. Networked writing involves determining, filtering, and forging connections between information and bodies. But where do these practices meet? Where is the transformation zone on the body that allows these practices to merge? The first obvious place to look is the location where human body meets the outside world: skin.

SKIN AS LOCAL AREA NETWORK

The connective membrane that encapsulates our singular bodies is called skin. Like Taylor, Claudia Benthien points out that early on in life, we see the skin as the boundary between ourselves and the world: "It is through the skin that the newborn learns where she begins and ends, where the boundaries of her self are" (7). Interestingly enough, skin and brain are formed from the same biological substance: "In the embryo, the skin and the brain are formed from the same membrane, the ectoderm; both are, in essence, surfaces. A substantial part of embryonal perceptions in the womb occurs through the skin" (Benthien 7). While skin and brain are both formed from the same membrane, and thus retain a certain membrane quality, they are also, in Taylor's language, types of screens or filters for information. The information we screen may come to us in terms of bacteria or "germs," but it may also come to us in terms of linguistic structures and cultural patterns. In fact, the viruses we encounter as human bodies and the viruses we encounter online may not be that much different. Thus, while our skin processes bacteria, it also processes and reflects cultural norms and patterns of meaning making; both natural and cultural processes are happening simultaneously. It makes sense, then, that in networked structures we use "skins" to both personalize and process the interactions between ourselves and others. We can first easily theorize skins through the interactions between ourselves and machines.

In online contexts, skin has two ostensible meanings, which are significant to an understanding of network culture. First, skin refers to the graphical avatars used in gaming communities; sometimes these avatars merely reflect the cultural influences and rituals of the game. However, in other instances, these skins can actually challenge those influences. As Allyson Polsky notes in an examination of female gaming culture, "Female gamers began producing experimental skins that challenge real world gender ideals and the aesthetics of both their male peers and the male-dominated gaming companies." The avatar as skin emphasizes the visual representation of the self in embodied terms; avatars, though, are usually perceived as "stand-ins" or representations of real-life bodies. The second meaning of skin is also of interest here. This meaning refers to the graphical layer or covering that users create to personalize software interfaces. For instance, the desktop interface for one's MP3 player can be personalized using a "skin" that is downloaded and installed on one's computer. But what is the relationship between the digital skin that represents the embodied subject and the skin that covers or layers the technology? The relationship is much the same as we see in the purely physical world.

In basic physical terms, we also seek to customize our bodies through all kinds of body modifications. Body piercing, tattooing, and other bodily modifications such as cosmetic surgery, allow the human form to function as a screen on which to project our selves and form identities. Visual culture has allowed and encouraged the human body to become the new *tabula*. As Taylor theorizes, "Body art represents, among other things, a sustained effort to reverse the dematerialization of art by making the body matter" (*Hiding* 111). Tattooing, for example, allows the body to show up; it reveals to us that just like the brain, the skin also serves as cultural screen and processor. Skin is an important screen, a part of knowing, thinking, understanding and accessing the world. A new way to conceive skin, then, is as the body's local area network, which means as a confluence of nodes or access points through which meaning is processed and created. Local Area Networks, or LANs, are small, typically invisible networks, but their very invisibility allows for other things to show up. As Taylor explains, "In many ways, the transformation of reality by electronic media is a practical extension of processes of abstraction and dematerialization (pre)figured in high modern art. When the body appears by disappearing on the screen, it becomes effectively material" (*Hiding* 111).

We know that our bodies effectively "disappear" when we are positioned in front of our networked computer screens.[4] But even if our physical bodies appear absent, they show up in all kinds of linguistic and social processes that occur in online interactions and publications. While we do have ways of describing and presenting bodies online (avatars, images, webcams), those forms are usually considered stand-ins for actual bodies in front of computers. If, as Taylor writes, the "self is the result rather than the presupposition of screening information," then subjectivity is clearly always in process and cannot be accounted for by a static visual object (*Moment* 205). Skin as visual depiction and skin as membrane would seem to be incompatible definitions, but according to Taylor, this would be part of network logic. The skin is screen, composed of nodes that are networked in both visible and porous ways. Thus, we can see the skin as screen, but because it is porous, it is constantly shifting, in motion, taking in information and exuding it. This is why body marks constantly shift in meaning depending on informational influences and contexts. But what of the relationship, then, between our computational screening of information over online networks and the LAN that is our skin? What is the relationship between the body of the user and the connections to information and others created online?

One of the reasons why many theories of embodied rhetoric have not addressed technology might be that technology poses a threat to the perceived organic unity of the body. It exposes the disruptive and refracted nature of the user in front of the computer screen. It also exposes the body as information streams. Consider the proliferation of the "skin" movement online: hundreds of individuals have created thousands of "skins"—digital coverings—for their software interfaces. In this respect, the interface is the prosthesis through which physical and digital bodies merge. How are we to distinguish between an online skin and a material skin? As Taylor explains, "In network culture, technology is an indispensable prosthesis through which body and mind expand. This relationship is always two-way: as body and mind extrude into world, world intrudes into body and mind" (*Moment* 231). This is why technology, and in particular the technology of the computer, provides a means for us to understand the ways that networks operate.

BROADBANDING BODIES

While we each have skins and we attempt to (re)create the permeable character of that skin online, we also must think about the ways that human bodies become permeable screens on a broad scale. One particular form that allows us to consider this idea is the webcam. More popular in the earlier days of the Internet explosion, webcams allows a user to hook up a mini-camera to a computer and stream a continuous camera image to a particular web page. The image is usually refreshed every few minutes, giving the impression of "live" images. Webcams are utilized to project a variety of images, such as views of a college campus, views from an office, or more infamously, views of coffeepots in break rooms. Most people, though, use webcams to project images of themselves for family and friends to view, often as they are working at their computers.[5] Because the webcam projects an image of the operator, the webcam is a useful form for thinking about the physical body in network culture and the complexity of the body's relation to the screen and ultimately to the network.

The image of the webcam operator staring into her screen on your screen is one of the most familiar images of the webcam movement. Michele White notes that in the webcam genre, "Binary distinctions between operator and spectator are disturbed by the way that the webcam mirrors the spectator's bodily position in front or the screen. [. . .] The spectator is already intimately close to the computer or even too close to see" ("Between"). Since most webcams are positioned next to the operator's computer, the spectator or viewer's position in front of the computer is mirrored in the webcam operator's position. In this instance, the screen simultaneously captures and conflates the spectator's and the operator's positions, making it difficult to determine who is spectator and who is operator. White explains that often, computer spectators become so wrapped up in the image that they are unable to grasp the whole representation ("Too Close" 20).

Moreover, as White observes, depictions of eyes on many webcam sites reinforce this confusion with seeing on screen. An example comes from the website JenniCam, one of the first and most famous webcam sites. In the image heading at the top of her main page, Jennifer Ringley, operator of JenniCam, is featured holding the web camera up to her eye. She seems to be looking through the webcam at the visitors to her site. As White claims, the webcam apparatus seems to supplant one of her eyes and "A fuzzy depiction that is reflected in the camera lens appears to be the body of the spectator" ("Between"). But the depiction in the lens is not clearly the spectator's body. The Rorschach-type image could also be the reflection of the computer screen or the reflection of the operator's body on the screen. The uncertainty of the image in the lens attests to the multiplicity of screenings that are happening on the network.

Further complicating webcam screenings are images of Ringley and other webcam operators wearing glasses. White reminds us that

> Glasses may screen the computer operator's eyes from the viewer while also allowing her to see. Seeing the screen reflected back in the operator's glasses reminds spectators of their own physical relationship to the computer screen. Less detailed reflections may even suggest that the spectator is reflected in the operator's glasses. Glasses impose a mediating frame, a kind of screen, between self and other. ("Between")

These on-screen examples witness the body of the spectator becoming another information process, just as the body of the operator becomes an information process caught up in the transmittal of her body into bits and bytes. The body becomes image, simultaneously captured and transmitted as data. The body's position in relation to the screen and the adaptability of screening the body through mobile computing devices further confuses the spectator with the screen. "In this sense," White explains, "all computer spectators become collapsed with the computer and may fail to distinguish where subject ends and object begins" ("Too Close" 20). The spectator is no longer a separate subject who processes or views the image; rather, the spectator becomes part of the screen, "literally mirrored, doubled, and confused with the screen" (23). The transmission of pictures/images that people take transfers the individual's vision to the vision of the machine.

This body as screen that happens on the webcam network is similar to what happens in some pieces of hypertext fiction. N. Katherine Hayles describes the net/work that is performed in Talan Memmott's *From Lexia to Perplexia*, a piece of hypertext fiction that challenges us with the realities of network culture. Hayles characterizes computers as simulation machines that produce environments:

> To construct an environment is, of course, to anticipate and structure the user's interaction with it and in this sense to construct the user as well as the interface. When the simulated environment takes literary and narrative form, potential possibilities arise for reflexive loops that present the user with an imaginative fictional world while simultaneously engaging her with a range of sensory inputs that structure bodily interactions to reinforce, resist, or otherwise interact with the cognitive creation of the imagined world. (48)

Memmott takes advantage of these flexible loops to structure the viewer's experience so as "to understand herself as a permeable membrane through which information flows" (Hayles 50).[6] Using what Hayles terms a "creole" language comprised of English and computer code, Memmott attempts to force viewers to see both our connections and our disjunctions with computers as reading and writing technologies. At one point in the work, Memmott carefully layers text and meaning three-foldly, so as to make the viewer think that "the self is generated through a reflection on the inside of the screen, as if "on the inside of a mask" (Hayles 52). Across another layer, "the subject and the techno-object are both inside, interfaced with the world through a screen that functions at once as display and reflecting surface" (52). And finally, a doubling third layer that "positions us inside the screen as well as external to it" (53). Hayles sees this movement as forcing viewers to recognize that their bodies are also undergoing transformations in reading the text, that they too are part of the cybernetic circuit (52).

In both the webcam genre and in Hayles' reading of Memmott's work, it becomes clear that the body functions as both a screen and a screening that is always already networked with other processes. In Taylor's terms, the connections between viewer and computer illustrate the networked processes of a complex, co-adaptive system. But while the model for this system in these examples is based on the individual viewer and computer, it can also be extrapolated to complex systems among many individuals. Recently there has been a proliferation of social software that allows many conversations and discussions. Friendster, Orkut, and other online "communities" attempt to connect or "network" individuals through their software. Individuals are depicted through digital images, providing the illusion that the image represents an authentic body. One of the reasons why these pieces of networked software ultimately fail is because they attempt to represent physical bodily patterns online. Clay Shirky explains, "A group of people interacting with one another will exhibit behaviors that cannot be predicted by examining individuals in isolation" ("Social Software"). Therefore, social software cannot attempt to replicate individual human usage patterns because those patterns will constantly change and alter; they are nomadic processes.

BODILY ASSEMBLAGES

The nomadic processes of groups emerge from the complex co-adaptive system formed among human bodies and computers. The coupling that makes possible these networks is the connection between body and machine—what we might term an assemblage. Deleuze describes an assemblage as being "populated by becomings and intensities, by intensive circulations, by various multiplicities (packs, masses, species, races, populations, tribes [. . .])" (Boundas 105). The assemblage relies on the structure of the rhizome, a concept based on principles of connection and heterogeneity, territorialization and reterritorialization. In *A Thousand Plateaus*, Deleuze and Guattari liken the rhizome to the relationship between wasp and orchid: "The orchid deterritorializes by forming an image, a tracing of a wasp; but the wasp reterritorializes on that image. The wasp is nevertheless deterritorialized, becoming a piece in the orchid's reproductive apparatus. But it reterritorializes the orchid by transporting its pollen" (10). But the concept of assemblage has a dual nature: "On the one hand, it is a *machinic assemblage* of bodies, of actions and passions, an intermingling of bodies reacting to one another; on the other hand it is a *collective assemblage of enunciation*, of acts and statements, of incorporeal transformations attributed to bodies" (88). They go on to say that "the material or machinic aspects of an assemblage relates not to the production of goods but rather to a precise state of intermingling of bodies in a society, including all the attractions and repulsions, sympathies and antipathies, alterations, amalgamations, penetrations, and expansions that affect bodies of all kinds in their relations with one another" (90).

One clear place we see these assemblages in action is through "smart mobs," Howard Rheingold's term for the phenomenon of people acting in concert regardless of whether they know one another: "The people who make up smart mobs cooperate in ways never before possible because they carry devices that possess both communication and computing capabilities" (xii). Smart mobs are made possible through mobile phones, laptops, PDAs, or any device that allows for communicative coordination between individuals. This coordination is characterized in physical terms, meaning that it is often used for physical aggregation of bodies in a particular time or place.[7]

Rheingold refers to the concept of this physical coordination as a mobile ad hoc social network, or

> the new social form made possible by the combination of communication, computation, reputation, and location awareness. The *mobile* effect is already self-evident to urbanites who see the early effects of mobile phones and SMS. *Ad hoc* means that the organizing among people and their devices is done on the fly, the way texting youth everywhere coordinate meetings after school. *Social network* means that every individual in a smart mob is a "node" in the jargon of social network analysis, with social "links" (channels of communication and social bonds) to other individuals. (169–70)

Derived from smart mobs, these ad hoc mobile networks enable human bodies to spontaneously assemble in a physical place through the mobile connections. He describes several scenarios where people have assembled for political action through mobile ad hoc social networks. For instance, he writes, "On November 30, 1999 autonomous but internetworked squads of demonstrators protesting the meeting of the World Trade Organization used 'swarming' tactics, mobile phones, websites, laptops, and handheld computers to win the 'Battle of Seattle'" (158). Rheingold sees the potential for political action through the networking of bodies via machines and mobile communications. But the network is not just reliant on mobile networks for its potential; it is also reliant on the assemblage of the human and technology. Rheingold incorrectly explains that it is because we *carry* these devices. I would argue that it's not a matter of carrying the device; rather, we have become inextricably connected to the device, creating a machinic assemblage. Cell phones, pagers, and other mobile computing devices are no longer seen as amenities but as necessities for living. The assemblage that is formed from human and cell phone (or other mobile computing device) allows those physical connections to take place through invisible network transmissions. Thus, the invisible networks that allow the skin to process and reflect information, the neural processes that form the nexus of our brains, are extended to the invisible wireless and microwave transmissions of mobile computing devices.

CONCLUSION: AFFECT

At this point, it might seem easy to connect these ideas with Brian Massumi's theory of affect as outlined in *Parables for the Virtual*. In that text, he conceptualizes the body beyond mere materialism and toward movement: "to think the body in movement means accepting the paradox that there is an incorporeal dimension *of the body*. Of it, but not it. Real, material, but incorporeal" (Massumi 5). Massumi describes this dimension as affect, or "a body's *capacity* to enter into relations of movement and rest" (15). Affect might be described as that intangible ability of the body to produce change while itself having the ability to change in response to bodies/affects. How is this different from Taylor's theories of emergence and network culture? Massumi and Taylor use different vocabularies to describe similar phenomena. Both want to move beyond the nature/culture binary and account for ways to understand bodily selves and produce material change. The difference I see is in the application of their theories. Massumi specifically moves against application: "The first rule of thumb if you want to invent or reinvent concepts is simple: don't apply them" (17); in contradistinction, Taylor spends an entire chapter outlining a plan for application ("The Currency of Education"). I think we can mediate between these two poles with our examination of how certain practices of screening have emerged, especially in Internet culture. Such an examination would not be for the purposes of application, but perhaps to theorize how we may not be able to apply complexity theory to Internet environments. This does not mean that I would privilege Massumi's paradigm over Taylor's; rather, while I find Taylor's theories conducive for theorizing Internet environments, they cannot fully account for the ways the individuals will use machinic assemblages in connective ways. Most of the ways that people have adapted technologies for their own uses have not been predicted or theorized beforehand; this is because these adaptations are always changing and altering with the (re)circulation of data and information.

Moreover, while Massumi's theory of affect does account for embodied connections between individuals, it does not fully explain the linking and transmission of bodies through wireless communications. As wireless signals and transmission hubs become more pervasive, our bodies will be constantly moving through wireless transmissions all of the time; how will our bodies interact with the complex co-adaptive wireless transmission system? Sometimes, when the cable signal on our television is out, I can use my body as a kind of antenna to help the television signal come in. This makes me wonder how our bodies will function within the complex evolving system of wireless transmissions. While our bodies constantly move through information and form the screen for that information, while simultaneously, in looplike fashion, processing and screening it, affect as we might know it will change. Wireless signals and electronic transmission will merge with bodies and transpose affect into a biotech phenomenon, if it isn't already.

NOTES

1. In composition studies, there are several works from the 1980s that discuss the shift to reading and writing on computer screens, but this research has been

mostly left dormant. For example, see Haas. In rhetorical theory, both Kenneth Burke and James A. Berlin speak of "terministic screens" in their work. However, as it is used in rhetoric and composition, "terministic screen" often lacks attention to the material screens through which we currently write and communicate. See Burke and Berlin. Blakesleys *The Terministic Screen: Rhetorical Perspectives on Film*, alludes to the terministic screen as a material screen with reflective properties.

2. I use the plural for embodied rhetorics to indicate that there are many different ways of conceiving an "embodied" rhetoric, and, in actuality, the term is not defined uniformly. For some, an embodied rhetoric might refer to Crowley's review of recent theories of the body and her call to move beyond the modernist subject, and Hindemann calls her attention to the teacher's body an "embodied rhetoric." I would align my approach with Crowley's.

3. The most common form of machinic connection in the West is the personal computer or PDA, but for other examples we could also look at the thousands of Asian workers who form complex co-adaptive systems with the motherboards their bodies create, and so on.

4. At the same time that our bodies seem to disappear on screen and in online interactions, we also become much more aware of our bodies through the effects of typing and sitting for long periods of time. While there has been a great deal of medical research on the potential detrimental effects of computer use, no clear conclusions have been drawn.

5. There has been some research on the webcam form, but not on the rhetorical principles of the form. Michele White, whose article "Too Close to See," is cited later in this essay, has done some of the most extensive work on this genre. See also Theresa Senft's earlier work on the webcam form.

6. Hayles vacillates between using the term "reader" and "user" in her discussion of Memmott's audience. I prefer to use the term "viewer," which reinforces the comparison with the webcam example and reinforces the focus on the image.

7. Smart mobs do not necessarily require a group of individuals to gather in one specific place. They can also refer to the phenomenon of moblogging, or mobile weblogging, where amateur or professional writer/journalists chronicle live events as they happen. Sometimes they upload these chronicles to the web or they may even video or audio record these events for others to see. In many cases, these accounts are performed to promote specifically political action.

WORKS CITED

Benthien, Claudia. *Skin: On the Cultural Border Between Self and World*. New York: Columbia UP, 2003.

Berlin, James A. *Rhetoric and Reality*. Carbondale: Southern Illinois UP, 1988.

Blakesley, David, ed. *The Terministic Screen: Rhetorical Perspectives on Film*. Carbondale: Southern Illinois UP, 2003.

Boundas, Constantin V., ed. *The Deleuze Reader*. New York: Columbia UP, 1993.

Burke, Kenneth. *Language as Symbolic Action*. Berkeley: U of California P, 1966.

Crowley, Sharon. "Body Studies in Rhetoric and Composition." *Rhetoric and Composition as Intellectual Work*. Ed. Gary Olson. Carbondale: Southern Illinois UP, 2002.

Deleuze, Gilles, and Felix Guattari. *A Thousand Plateaus*. Minneapolis: U of Minnesota P, 1987.

Haas, Christina. "'Seeing It On The Screen Isn't Really Seeing It': Computer Writers' Reading Problems." *Critical Perspectives on Computers and Composition Instruction*. Ed. Gail E. Hawisher and Cynthia L. Selfe. New York: Teachers College P, 1989. 16-29.

Hayles, N. Katherine. *Writing Machines*. Cambridge: MIT P, 2002.

Hindemann, Jane E. "Writing an Important Body of Scholarship: A Proposal for an Embodied Rhetoric of Professional Practice." *JAC* 22 (2002): 93-118.

Lupton, Ellen, with essays by Jennifer Tobias, Alicia Imperiale, and Grace Jeffers. *Skin: Surface, Substance, and Design*. New York: Princeton Architectural P, 2002.

Polsky, Allyson D. "Skins, Patches, and Plug-ins: Becoming Woman in the New Gaming Culture." *Genders* 34 (2001). http://www.genders.org/g34/g34_polsky.txt (27 Feb. 2004).

Manovich, Lev. *The Language of New Media*. Cambridge, MA: MIT Press, 2001.

Massumi, Brian. *Parables for the Virtual*. Durham: Duke UP, 2002.

Rheingold, Howard. *Smart Mobs: The Next Social Revolution*. Cambridge: Perseus, 2002.

Selzer Jack, and Sharon Crowley, eds. *Rhetorical Bodies*. Madison: U of Wisconsin P, 1999.

Shirky, Clay. "Social Software and the Politics of Groups." 9 Mar. 2003. http://shirky.com/writings/group_politics.html (3 Sept. 2003).

Taylor, Mark C. *Hiding*. Chicago: U of Chicago P, 1997.

——. *The Moment of Complexity: Emerging Network Culture*. Chicago: U of Chicago P, 2002.

White, Michele. "Between Domestic Technology and Empowered Visibility: Women, Webcams, and the Public Sphere." 2002 Rhetoric Society of America Conference. Rhetoric Society of America. Alexis Park Resort, Las Vegas. 23 May 2002.

——. "Too Close to See: Men, Women, and Webcams." *New Media and Society* 5.1 (2003): 7–28.

Young, Iris Marion. *"Throwing Like a Girl" and Other Essays in Feminist Philosophy and Social Theory*. Bloomington: Indiana UP, 1990.

3
OUR CYBERBODIES, OURSELVES
CONCEPTUAL GROUNDS FOR TEACHING COMMODITIES TO WRITE

Stanley D. Harrison

Looking into my computer classroom these days produces feelings ranging from vertigo to exhilaration. When I open the door, strange students are there to greet me, and I pause at the sight, tremble at the thought, and wonder at the fact of them—these posthuman students of mine.

Theta student, sitting at eir[1] computer workstation, has a phone pack on eir belt and a wireless headset on eir ear. Without breaking from work on eir writing assignment, e rings at the belt, presses a button on eir phone pack, says "Be home at six" and "Love you, too" at eir headset mouthpiece, and then turns off eir mouth by repressing eir phone pack belt button. Next to em, *Gamma* student senses my presence at the door and closes out of a well attended transvestite chat room. Momentarily chagrined, e tries to cover eir tracks by performing an Alta Vista search for keyword "Harvard." Alta Vista knows better than to trust *Gamma's* erudite impulse of the moment. The search engine processes both *Gamma's* ivy league search term and eir transvestite cookie before coming back with text links to Harvard University and animated graphical links to *Frederick's of Hollywood*. Rumor has it that *Gamma* waited until I took up position in front of the class before linking to *Frederick's* and purchasing the establishment's fabulous Harem Dancer Costume set—a four-piece ensemble that includes a veil, sequined headband, foamed shaped bra, and sheer pants with full back and built-in panty.

What a peculiar lot they are.

Small wonder then that I should feel as I do, like a teacher in a strange land peopled by creature students who profess to be human but who look and communicate like students from a distant planet. Still smaller wonder that the sight of *Alpha* student should fix me as e does, causing me to stare out at em from my place before the class, a fascinated witness to the act of SoundWriting. E, like *Theta*, speaks at eir headset mouthpiece, but, unlike *Theta*, eir voice is tethered to a computer by a short patch cord. When *Alpha* speaks, eir microphone transducer turns eir sound wave energy into a continuous flow of electrical energy that eir voice-to-text computer program changes into strings of interrelated but disconnected and rearrangable digital information, which eir computer motherboard both stores in short term memory and reconstitutes as written words on a video screen. In other words, when *Alpha* speaks, *Alpha* SoundWrites, making text appear on eir computer monitor to the tune of one-hundred and eighty words per minute—as e does when e quotes me in a paper e is writing on contemporary composing processes. "According to Stanley Harrison," e composes, "SoundWriting 'challenges the idea of speech as ephemeral activity, shifts the site of composition from hand to mouth, and increases the efficiency with which we produce written text. To be sure, because posthuman SoundWriters use a supernormal process to produce text in increasingly supernormal amounts, we might reasonably expect an intense, if not supernormal, debate to hinge upon the exploits and adventures of posthuman SoundWriters.'"

And then there is *Omega*. E looks at me from behind a pair of "smart eyeglasses," which e says connect to a computer e wears under eir shirt. E clicks a handheld control and manipulates a mouse rollerball. Of course, the mouse clicking and rolling might signify nothing, but *Omega* is probably either sending a picture of me via wireless eyeglass webcam to a remote display station or opening an e-mail that displays for reading on the underside of eir eyeglasses. In private conversation, *Omega* has claimed that eir tetherless system makes eir flesh body "smart": "The kind of synergy that arises from constant connectivity" is particularly strong, e says, "because [human-machine] interaction is sustained" over a long period of time (Mann). E says, by way of example, that wearing "smart eyeglasses" while grocery shopping has transformed the way eir eyes process fruits and vegetables. "I stare, let's say, at a cucumber display," *Omega* continues, "and, somewhere else, my wife looks through my eyes, inspects the produce, and e-mails me with comments" (Mann).[2]

Many times each class day, I look upon my uncanny students and see one incontrovertible fact: they are ceasing to be human beings in the traditional sense with increasing regularity. Whether they close quarters with hands-free cellphones, undergo subject position manufacture while surfing the cookied internet (see Johnson-Eilola, "Control"), forge intermittent but nonetheless protracted connections with their computers in order to produce SoundWriting, or merge with tetherless "smart clothing" computing systems, my students fall or rush into posthuman cyborg states before my very eyes. When they take up the positions provided for them in my institution's computer classroom, they do—at different times and to varying degrees—what all good cyborgs do: they become homeostatic systems functioning unconsciously (see Clynes and Kline). In other words, they abandon their humanity for the privilege and burden of having powers and pressures beyond those of mere mortals.

This, however, is not the end of things.

What I also see is that my students, in becoming cyborgs, accede to being nothing less than "living commodities" in the literal sense of these words; that is, they put on their prostheses and, in so doing, transform themselves into manufactured, animate, disposable exchange values that must pay to maintain and/or upgrade "themselves" if they are to survive as cyborgs. *Theta* student, for example, becomes just another human being who talks on the telephone if e forgets to pay eir monthly cell phone bill. To be sure, what animates the living cyborg—the software or public utility that fuels the cyborg's prosthesis—comes to us in the form of a ticket item that seems only to fuse with the cyborg's flesh if the cyborg, or the cyborg's patron (for example, a college with an open computer lab), agrees to pay or, alternatively, to enter into an arrangement with capitalists for deferred payment. Indeed, a subtle arrangement for deferred payment accounts for the continued existence of *Gamma* student, my web-surfer in the harem outfit, whose free-linking, hypertextual subject position gets appropriated by capitalists who, for their part, use such things as Internet cookies to bind *Gamma*'s hyperlink movements to products, purchases, and consumption (see Johnson-Eilola). The cost of *Gamma*'s web surfing is regular exposure to user-triggered, user-specific hucksterism that theoretically results in a user-defined, user-purchased range of products and services.

When it comes to *Alpha* and *Omega*, the cyborg imperative to purchase the commodity self becomes intensified to the point of explosion. Both tethered and tetherless human/computer homeostatic systems transform unsuspecting computer users into network-ready, software-driven, hardware-supported, biological workstations. Even more, human/computer systems shift the electronic contact zone from the computer screen interface to the software-driven, hardware-supported nervous systems of biological workstations; undergo regularly scheduled technological obsolescence; maintain their embodied cultural identity through a program of software and hardware upgrades; and, from the outset, exist as actual commodities that know they must continue to purchase themselves if they would sustain themselves as commodities with identities. Seen from this perspective, *Alpha* and *Omega* exemplify, with one exception, what Donna Haraway meant when she wrote, "Cyborg writing is about the power to survive, not on the basis of original innocence, but on the basis of seizing the tools to mark the world that marked them as other" (94). All I would add is that the power to survive, which is at the heart of cyborg writing, seems less about "seizing the tools" than about paying for the programs and program upgrades that constitute both the cyborg's world (for example, Windows 3.1, Windows 95, Windows 98, Windows ME) and the cyborg's identity (Dragon NaturallySpeaking, Microsoft Office Professional Edition). Simply put, *Alpha* and *Omega* must cease to exist if they ever lack money to pay for those components of their cyberselves that have a quantifiable exchange value.

Strange, indeed, to look upon beings that are both the commodities and end-users of a system that wants its posthuman products to think of themselves as human users and not as commodities. Strange, also, that I should be so affected by the sight of these living curiosities. They are still my students, after all. But something has changed, something fundamental, and I find myself grasping in vain for the conceptual tools that will, on the one hand, get me past my desire to walk into a classroom and see it populated by human computer users and, on the other, get me on with the task of helping these student commodities to mark the world that marked them as other. This much is certain: at a time when I need to acknowledge, if not appreciate, my students for the commodities they are, the open literature on computers and composition threatens to set in motion an unproductive nostalgia in need of correction—a nostalgia for bygone, if not illusory, days when teachers fought for the rights of "human" program-users, not posthuman "student-programs."

Accordingly, my purpose here is to argue for a pedagogy of the posthuman that more completely meets the needs of those student writers we encounter in computer classrooms. Toward this end, I shall point out that the educational theorists of interface do offer critical approaches to the problem of computer writing, but they skew their proffered critical perspective by drawing impossible lines between human agents and their computerized instruments and environments when they suggest that critique of computer interface originates from points "outside" of technology or the human-computer connection. This perspective, my argument holds, contributes to an unproductive nostalgia for "the human" because it supports composition theorists who would "pay attention to technology" in order to become better "humanists" and serves as an advocate for the technological underclasses, not for cyborg theorists who would instruct writers faced with the challenge of becoming posthuman cyborgs at the point of interface.

For the purposes of correcting the "humanist" error in critical theories of interface, my essay will introduce the following proposition: we fail to serve the needs of posthuman students whose subjectivities *emerge at* the conjuncture of consumer culture and cyberspace and *emerge as* commodities (that is, as disposable market, as opposed to human, values) when we apply what Joseba Gabilondo calls the ideology of "Man" to the cyborg problematic—that is, when we come at the problem of computer-generated cyborg existence through the phallogocentric understanding that First World nation-states produce the democratic, middle-class, consumer, "Man" subject position as their first order of business. The effect of this observation will be to support my claim that the sudden appearance of commodity students requires that we produce critique that both gauges the pressures and limits that define commodity, or cyborg, subjectivity and, just as importantly, resists the distorting influence of compositionist nostalgia for the human. Coming at the problem of cyborg writing from this vantage will reveal that cyborg writers, in the first instance, are born into the nonsociety of ignorant, self-involved cyborg writers whose first order of business is to purchase, master, dispose of, and upgrade their prosthetic selves—in other words, to acclimate themselves to a commodity-driven, blissful tyranny of the subject-self over the object-self that exists to be bought, used, and destroyed at the point of upgrade. With this in mind, I will argue that critically oriented compositionists must accept the power of consumer culture and cyberculture to transform human life into a commodity fetish, relinquish the ideology of "Man" when faced with the task of teaching cyborg writers, and embrace posthuman critical pedagogy in hopes of demystifying cyborg-filled computer classrooms and, more importantly, designing curricula suited to the needs of our posthuman students. Consequently, we need to begin immediately to do such things as teach student cyborg writers how to intervene in their subject formation at the level of software so that they might learn to participate in the counter-hegemonic manufacture of a cyborg self that is not, at one and the same time, a living commodity.

HUMANIST THEORISTS

To their credit, theorists like Joel Haefner, Johndan Johnson-Eilola, Cynthia Selfe, and Richard Selfe bring a critical attitude to bear on the problem of interface. As a group, they stress, for example, that subjectivist individualism (the idea that computer users shape their environments according to their creative will) is inadequate to the task of accounting for language origins and practices in an era dominated by politically articulated computer writing spaces. Toward this end, they recommend that teachers identify the cultural metaphors and premises that shape the computer interface and the computer-user's experience, believing that close analysis of the computer interface will not only help educators perceive the effects of "domination and colonialism associated with computer use" but also empower educators to "establish a new discursive territory within which to understand the relationships between technology and education" (Selfe and Selfe 482). The educational theorists of interface also suggest that we take an active hand in customizing the programs we use in our composition classes because Structured Programming protocol—the heart of ubiquitous American computer program code, interfaces, and operating systems—is itself shaped by what Haefner calls "the profiteering imperative and the hierarchical structure of corporate America" (325). Finally, these theorists express concern about our automatic preference for first-person and argumentative essays, as well as literature and literary criticism, arguing that this preference blinds us to the power of functional hypertexts (online help, for example) to underwrite composing practices that value transparency, efficiency, and performativity over contingency, dissensus, and negotiation (Johnson-Eilola, *Nostalgic*). As a group, the theorists of interface advocate politically oriented critical literacy of computer technology, as opposed to task-oriented functional literacy, and they seem entirely unmoved by nostalgia for challenges associated with teaching in the pre-computer age.

At the same time, however, these theorists lose their edge where they succumb to an uncritical, ultimately disempowering nostalgia for a "humanity" that not only exists "outside of technology" but that also gains perspective on and effectively alters key points "inside of technology." While they are correct to protect against excesses in the direction of subjectivist individualism, the theorists of interface, interpellated by the ideology of "Man," make an understandable albeit unfortunate error in guarding against impulses that run in the opposite direction—that is, toward theorizing that stresses the power of commercially organized and proliferated computer writing spaces to penetrate and utterly transform the computer user at the point of interface. Because educational theorists of interface need to believe that writers, writing teachers, and writing theorists should "*use the technology* to question the hegemonic tendencies of disciplinarity and discourse communities" (Johnson-Eilola, *Nostaligic* 28; emphasis added), they permit themselves to draw impossible lines between human agents and their computerized instruments and environments for the purpose of allowing persons to stand back from interface, to read interface critically, and to reengage with interface from the position of critical literacy and with the effect of altering the political trajectory of interface. Toward this end, they deploy the "interface as contact zone" metaphor with a certain frequency, and they dream of opening and privileging a nonexistent *outside* postcolonial space from which to bring the ideology of "Man" to bear on the *inside* "Cyborg" problematic.

Yet, the truth is with cyborg theorist Joseba Gabilondo, who observes that "there is no such thing/subject as a 'postcolonial cyborg,' because postcolonial subject positions are *always left outside cyberspace*" (424; emphasis added). By way of explanation, he writes that "the production of 'Man' [in the economically privileged First World] has given way to the reproduction and simulation of 'cyborgs,' and the technologies and apparatuses of the nation-state that produce the democratic, middle-class, consuming 'Man' have been transferred to the peripheries of the First World and to the Third World" (424). From this perspective, members of the set "Man" are either present in the Third World, where access to computers is a chimera; or prevalent on the peripheries of the First World, where access to computers is restricted or denied; but they are never members of the set "Cyborg," which includes no members of the set "Man," because interface transforms persons into cyborgs. To insist that human beings are not fundamentally transformed at points of human-computer conjuncture misses the point of interface: intimacy with computers takes hold of fleshy beings, typically born into this world as use-values who spend their lives fending off cultural pressures to become exchange-values, and changes them into cyborgs, borne by interface into a state of being an exchange-value that might, through sustained effort and cunning, become and then die as a use-value.

Bringing the ideology of "Man" to bear on the "Cyborg" problematic produces some rather interesting effects on behalf of the technological underclasses but contributes nothing to a philosophy that would help cyborgs (e)merge with(in) a giving, sympathetic, and self-controlled society of cyborgs that both values its citizens and, also, treats "Man," or being-prior-to-interface, with respect. This is what comes across when we read, from a heretofore undefined cyborg perspective, articles like Cynthia Selfe's "Technology and Literacy: A Story about the Perils of Not Paying Attention." She appeals to the ideology of "Man" where she argues that the fight against the continuation of racism and poverty through the unequal distribution of technology is a battleground for *humanists*. The problem, she makes clear, is that "in our educational system, and in the culture that this system reflects, computers *continue to be distributed differentially along the related axes of race and socioeconomic status* and this distribution contributes to ongoing patterns of racism and to the continuation of poverty" (420). The solution to the problem, which asks compositionists to pay attention to technology, is slow in coming because technology is "either boring or frightening to most *humanists*" (412; emphasis added). She suggests, shortly thereafter, that as *humanists* we prefer things to be arranged so that we don't have to pay attention to machines because "computer technology, when it is too much in our face (as an unfamiliar technology generally is), can suggest a kind of cultural strangeness that is off-putting" (413). Nonetheless, she believes that compositionists must take it upon themselves to merge "the technological and the *humanist* perspectives" and, in so doing, empower themselves to advocate "free access to computers for citizens at the poverty level and citizens of color" (434; emphasis added). By paying attention to technology, she concludes, we may "learn lessons about becoming better *humanists,* as well" (435; emphasis added).

Her perspective speaks to computers and composition scholars like Jeffery Grabill who argue, for example, that compositionists need to "work on access in nonschool settings" in order to prevent "the technopoor" from "missing something" (313). Clearly, this inside/outside approach to cyborg writing champions the cause of "Man" on the peripheries of the First World. Bearing this in mind, we need to acknowledge that the critical, political program advocated by the likes of Grabill, Johnson-Eilola, Haefner, Selfe and Selfe proceeds from the mistaken assumption that postcolonial "humanism" is consonant with efforts to educate posthuman cyborgs. We need to accept that cyborg students, like their human counterparts, need the help of teachers who will address themselves to the particular needs of their students. Cyborgs materialize for the duration of interface and demand an education appropriate to the needs of cyborgs, even if this means that their teachers fail to advance the cause of "Man." Only when interface, or techno-human fusion, is broken and "Man," with the sense memory of "Cyborg," reappears does the need for a pedagogy of the human reappear. These appear to be the "facts" and, as such, serve as a warning against those who would succumb to the all-too-human practice of treating cyborgs as human correlatives and humanizing educational protocols as a matter of unstated policy.

COMMODITY STUDENTS

But how do we make this change? How do we learn to see our students for what they are? How do we develop appropriate strategies for teaching the cyborgs that increasingly populate our classrooms? In the first place, we need to understand why pedagogy that is steeped in race, gender, and class analysis but that is not also grounded in class-inflected cyborg analysis must fail students who require a pedagogy of the posthuman. Toward this end, we need to accept that the experiential categories race, class, and gender correspond to a mode of production that manufactures human, as opposed to posthuman, subjectivity as its primary order of business. In *The German Ideology*, Marx and Engels provide an overview of the process that produces human subjectivity. The isolated human body, imagined for the moment as existing outside of culture, has needs. The satisfaction of the body's first need leads to new needs and, as such, constitutes the first historical act. "The third circumstance," Marx and Engels write, "which, from the very outset, enters into historical development, is that men, who daily remake their own life, begin to make other men" (49). Labor and procreation, or the production of life, result in the first mode of production, for bodies in collective are quick to discover the necessity of co-operation. The need to improve co-operation through communication, of course, is what leads to language acquisition, what Marx and Engels call "practical consciousness," or language that "only arises from the need, the necessity, of intercourse with other men" (51). Of obvious significance, human subjectivity emerges at this juncture as the result of the several divisions of labor: sexual labor (gender), physical and mental labor (class), and cultural labor (race). Indeed, the experiential categories race, class, and gender, which are the products of low- or no-tech embodied human interaction, are bound inextricably to that state of practical consciousness that Gabilondo calls "Man."

In contrast, the cyborg subjectivity that shows up in computer writing centers appears only at the technologically advanced stage of production when it becomes possible for human beings—as the products of established systems of race, class, and gender identification—to co-operate in the production, distribution, and reproduction of "intelligence amplifying" prostheses that both network the body and inscribe the body with semiotic traces of race, class, and gender distinctions that exist for cyborgs as powerful elements of the commodity prosthetic, as opposed to the product of lived human relations. In other words, culturally articulated human subjects decide to wear prostheses that have the effect of birthing people out of their humanity and into a superstructural zone that conflates the forces and relations of production in the cyborg prostheses and typically sells these prostheses on the open market, indicating that cyborg subjectivity materializes at the conjuncture of cyberspace and consumer culture.[3] That is to say, cyborg subjectivity begins when the self—or purchasing agent for the self—buys the tools that are to become indistinguishable from self and, in so doing, adopts a proprietary attitude toward the self as chattel—the logical result of a cultural system of production that encourages cyborgs to sell, buy, and become the self that *is* a market value and to have no moral regard for or sustained relation to the self, which becomes garbage at the moment of upgrade, the refuse of a life manufactured, sold, bought, and discarded.

Strictly speaking, race, class, and gender analyses, each based on a division of human labor, will fail to penetrate the cyborg subjectivity and, therefore, fail to produce a posthuman pedagogy. These modes of critical analysis, and others, will yet prove indispensable to the cyborg scholar, critic, and teacher because cyberspace is already fully raced, classed, gendered, aged, nationed, and so on. Yet, if our tools of analysis are to produce critique that escapes the pull of nostalgia for the human, they need to be rearticulated to address the problem represented by embodied subjectivities that are both mortal and commodity, both relation of production and material of production, both subject and object. For this to happen, we need to add at least one more element to our analysis of cyborg subjectivity as a subject/object conflation. We must ask ourselves, "What are the pressures and limits that define the subject/object subjectivity—including its potential for sustained, collective, counter-hegemonic action—when it appears as a consumable sign within the sign system of objects that manufacturers and advertising agencies produce in order to stimulate and control sales of the disposable self to the disposable self?"

Because interface transforms human beings, under typical circumstances, into self-consuming commodities, we need the help of a philosopher of consumption who is guilty of the kind of totalizing, deterministic, antisocial theorizing that holds small appeal for computers and composition theorists and does, in fact, overstep its bounds when applied to human culture. Jean Baudrillard is particularly useful here, even though his theories of hyper-reality, which depend on such ideas as "the implosion of the social," become unstable when applied to human communities. Statements to the effect that resistance to advertising is futile in a world where the social is a simulation crumble when we remember that human beings still participate, however minimally, in locally generated communal relationships (primary groups like family, church, school, and community watch) that deflect or inflect the influences of advertising on individual subjectivity. However, Baudrillard appears to have much to say to teachers of computer-based writing, when we consider that a commercial relationship between and for the continued existence of animate products results in cyborg subjectivity, and that cyborg subjectivity is consonant with commercial influence because cyborgs are borne by interface into direct relations with "providers," not primary social groups, who/that exist to deliver, not deflect, one message: become, dispose, and upgrade. For us, there is no resisting Baudrillard, not when newly self-purchased cyborgs come packaged to accept "the providers message"—that is, when they enter the human-computer world in a state of ignorant, self-involved isolation that amounts to an absence of relation; when they emerge as first-time cyborgs who do not know how to turn themselves on or off, let alone how to operate an e-mail client, access and establish Usenet newsgroups, participate in and host IRC, or contribute to and administrate W3 bulletin boards. For their part, cyborgs materialize as inefficient users inside an absent social order, a nonsociety of cyborgs, and this renders them, even as it leaves them, unprotected from the sale of themselves to themselves.

Because the cyborg depends upon self-consumption for its existence and is, therefore, vulnerable to the dictates of consumer society, we need to ask with Baudrillard, What is the experience of life within consumer culture at its most extreme? To answer this question is to see the cyborg's soul, and to know that posthuman pedagogy must make provision for teaching subjects that begin by being marked as other, even from and to themselves.

Many people still believe that the words "consumer society" refer to a society of consumers who participate in a self-directed activity of commodity consumption. Baudrillard, however, argues convincingly that "the ideology of competition, which . . . was previously the golden rule of production, has now been transferred entirely to the domain of consumption" (11). "A fixed class of 'normal' consumers," he continues, "has been created that coincides with the whole population," and, as importantly, capitalists have developed a strategy for controlling these consumers that includes materializing the superego; stimulating the id, or deep drives; and sanctioning/censuring consumers to act "freely" on their deep drives, or desires, in order to be different from everyone else through consumption and exactly like everyone else through consumption ("I ran Windows 98. Now I run Windows ME." "I ran OS. Now I run OS/2 Warp." "I am different. I am the same.").

The key here is in the materializing of the superego. Ordinarily, the superego is immaterial and exists because individuals participate in its production through syntactic linguistic exchange (speech acts) with members of a shared community. These syntactic exchanges give rise to, among other things, the superego, or unconscious consciousness, that may be defined in part as the internalized set of asyntactic expressions that defines (enables/limits) what speakers might say or think comfortably at any given time. The materialization of the superego through advertising subverts this process by providing consumers with a set of asyntactic expressions (Pentium I, Dreamweaver HTML editor, DSL) that emerges without the participation of the consumer. Because the consumer does not produce this set of expressions through syntactical exchanges, the materialized superego is always inappropriate to and incapable of integrating with the self. More importantly, the available set of expressions, which corresponds to related sets of products and recommended feelings, has no meaning for the consumer except insofar as it stimulates the consumer's desires (Chevy, Ram Tough—BuyIT) and then breeds dissatisfaction in the consumer in order to produce a new set of desires (~~Chevy~~ PT Cruiser, Retro, Cool—BuyIT). Rather than exalting intelligence and wisdom, the authors of this system seem to champion the "ideology of personal fulfillment," the "triumphant illogicality of drives cleansed of guilt"—that is, the regression by adults into a series of unrelated, albeit reproducible, infantile desires for and dissatisfactions with products (18).

The impact of consumer culture on cyborg subjectivity, which exists within advertising's system of salable object relations as a disposable exchange value, is both profound and unique to cyborg culture. Born into an absence of relation that is all about the purchasing of materials necessary to resolve a fundamental inadequacy in the self, the cyborg quickly learns that the self that will or should endure can never be bought. Cyborg subject positions are manufactured and sold with the intent of creating cyborgs who not only look upon the self as an object that must be re-consumed on a regular basis but, also, move into a relation of blissful tyranny over the self that exists to be bought, used, and destroyed. Cyberspace, given this arrangement, becomes something of a showroom display case filled with id-driven, self-involved, self-destructive subjectivities that are attractive to consumers because they are neither produced to forge meaningful, politically active communities on the Web nor produced to be self-aware of the fact that the battle for profits has expanded to include the cyborg's self-financed war upon the self.

Each is made to destroy, buy, destroy, buy, destroy the self, which exists to be different from but identical to other cyborgs who destroy, buy, destroy, buy, destroy the self.

POSTHUMAN PEDAGOGY

What precisely does this mean to one whose livelihood depends upon teaching cyborg writers to compose? If I am correct in my analysis of cyborg subjectivity, then compositionists will need to do more than consider the influence of computer-based writing tools and environments on the processes and practices of human literacy. We will need to go beyond thinking, for example, that the computer interface is a semiotic contact zone that privileges and empowers male, caucasian, American, corporate, human identity, even as it supports the creation of a technological underclass that includes disproportionate numbers of African Americans, women, and citizens of the Third World. Instead, computer writing specialists should move to understand, in the first place, that the mere fact of computer use renders the computer writer a cyborg, which is not merely a postmodern subjectivity but also the hegemonic, albeit self-destructive, subject position that orders cyberspace. Thereafter, they should embrace the fact that while "humanistic" composition research correctly registers that online experiences lead to the development of "*heterotopia*, spaces to be negotiated and transformed as a result of the conflict that arises within them" (Blair 318), and then inflects this understanding from the perspective of gender (Sullivan), sexual orientation (Comstock and Addison), race (Taylor), class (Whitaker and Hill), second language acquisition (Belcher), and physical disability (Buckley), "humanistic" compositionists cannot help commodity students address their cyborg-specific problems and create counter-hegemonic cyborg heterotopia without the aid of a posthuman pedagogy that stands upon this understanding: chances for radically democratic cyborg writing wane to the degree that cyborg writing spaces are populated by animate-product subjectivities that have yet to critique and rearticulate the cyborg problematic, or, life as the salable, self-destructive conflation of posthuman subject/object relations.

To be sure, even the gross particulars of this proposed posthuman pedagogy are unknown to us. Yet, the necessity of posthuman pedagogy for the improvement of commodity students requires us to speculate on the shape it might take, the directions it might lead. I suggest that computer writing specialists can take a meaningful step in the direction of posthuman pedagogy by opening the doors on their classrooms, looking in on their commodified cyber-students, and seeing that commodity students are flush with the desire to buy and destroy themselves, even as they are humiliated in this regard because they cannot spend the $2,313.80 it would cost to build, but not upgrade, a competent cyborg writer (see Table 1).

A shortened list of the writing tools that our students need "to become" before they leave college justifies the estimated cost and makes daunting Haraway's characterization of cyborg writing as survival on "the basis of seizing the tools to mark the world that marked them as other." Obviously, cyborg writers need to own and operate a word processing prosthesis (that is, a program) that saves text in the most widely supported word processing format (.doc). But can they afford this prosthetic device *and* the others they will need to complete themselves? They will want to use both mind mapping and tree outlining prostheses when developing and organizing their texts. They will also need a portable document file distiller (.pdf) so they can open and print files on any computer without producing changes in the document's original layout and design. Next, commodity students will require both a file compression and file splitting utility, for times when large files must either be shrunk down or split up and distributed over many disks. They will want to enhance their oral presentations with slideshows saved in the popular PowerPoint format (.ppt). Then, too, students will need to learn how to compose and maintain databases, if only so they can create and update a bibliographic database. Those serious about group writing will want to establish an Internet Relay Chat (IRC) room and use an IRC client to log group chat, send private messages, and exchange files in "real time." Being able to create virtual network interfaces (networking personal computers via the Internet) will prove helpful because writers on virtual networks can view and edit documents at the same time, connect to and write on home computers while on location, and provide direct technical support to writing group partners with computer troubles. Because students will want to produce help documents that will make computer documents and environments more accessible to the public, they will want to procure and learn to use a good HELP editor. All of our students should design and draft extensive academic Web sites that, on the one hand, comply with the current HTML standard and, on the other, support students in research and writing that happens while away from home; therefore, they will need to have and know how to use a high powered HTML editor with strong support for cascading style sheets. The need to program internet servers to accept HTML files and, thereafter, to upload files to the world wide Web makes a working knowledge of telnet and FTP clients essential. Of course, cyborg writers will want to send e-mail and participate in Usenet news groups, so they will need an e-mail client and newsreader. Increasingly pressured to conduct effective internet research, student commodities will want to acquire both a desktop searchbot and an offline browser, so they can query search engines and save their search results and, also, download entire Web sites for extended offline study. Finally, the frequency with which cyborg writers transfer files during group work necessitates that they procure reliable anti-virus protection.

By all outward appearances, the cost of becoming a competent cyborg writer exceeds the immediate grasp of most commodity students. And this is to the advantage of posthuman pedagogy, which demands that we exploit our students' inability to satisfy themselves through self-consumption, doing what we can to drive a wedge between the cyborg and the cyborg's consuming lust for self. We need to make our students aware, in their moment of financial

Table 1

Costs of Establishing Minimal Cyborg Writer Competence

Competency	Program	Cost
• MS Office Compatible Document Database, Spreadsheet, Slideshow, Desktop Publisher	Microsoft Office 2000	$499.00
• Idea Generator	Axon 2001	$160.00
• Outliner	Action Outline	$24.95
• HTML Editor with CSS support	DreamWeaver 3.0	$299.00
• Portable Document File Creator	Adobe Acrobat 4.0	$249.00
• Anti-Virus Protection	McAfee 5.1	$29.00
• IRC	mIRC	$20.00
• File Compressor	WinZip 8.0	$29.00
• File Splitter	File Splitter Deluxe 3.1	$11.95
• FTP	WS-FTP	$40.00
• Help Editor	RoboHELP Office 9.0	$899.00
• Offline Browser	Black Widow 4.07	$39.95
• Searchbot	Web-A-Matic	$12.00
• E-mail Client, Newsreader, Telnet	Outlook Express Newsreader and Windows Telnet	$0.00
Total		**2,313.80**

Source: CNET download.com <http://download.cnet.com/>, NoNags <http://nonags.gargantuan.com/index.html>, and TUCOWS <http://im1.tucows.com/>

weakness, that they need not purchase very much of themselves at all and, also, that cyborg writers may join with others of their kind in, for example, Usenet newsgroups that promote an alternative to commercial cyberspace. Our students must know that they can satisfy the cyborg's real need for software without activating the cyborg's infantile desire for disposable happiness through self-consumption. Indeed, compositionists who take time to become familiar with both Usenet freeware culture and the art of freeware self-fashioning—as opposed to pay, ad, and spyware self-fashioning—can advance their cyber-students toward this next understanding: living commodities should write their cyborg bodies with freeware software alternatives where possible, if only because this will help them to imagine and compose alternatives to the hegemonic subject position that the ideology of multinational capitalism privileges.

But how do we do this? How should we teach a living commodity to compose the self in opposition to the self? My immediate recommendation would be to create projects that force cyborg writers to do two things. First, they must confront their status as consumers who buy, use, and destroy the self in a never-ending cycle of self-sacrifice that has no purpose except to stimulate the self to buy the self. Second, students need to participate in newsgroups, like alt.comp.freeware, for the purpose of working with others to establish freeware collectives that will, among other things, satisfy the cyborg's real need for advanced writing programs.

Significantly, such projects will teach cyborg student writers to upgrade themselves at a cost to them of $0.00 (see Table 2). Even more importantly, such projects should help cyber-students to understand that the conjuncture of cyberspace and consumer culture manufactures self-consuming subject/object commodities, and not human beings; that cyborgs—cultural fictions that they are—cannot be made powerful by appeals, implied or stated, to the idea that

Table 2
Actual Costs of Establishing Minimal Cyborg Writer Competence

Competency	Program	Cost
• MS Office Compatible Document Database, Spreadsheet, Slideshow, Desktop Publisher	Open Office	$0.00
• Idea Generator	MindMan Personal	$0.00
• Outliner	KeyNote	$0.00
• HTML Editor	1st Page 2000	$0.00
• CSS Editor	Balthisar	$0.00
• Portable Document File Creator	GhostScript, Ghostview	$0.00
• Anti-Virus Protection	AVG Anti-Virus	$0.00
• IRC	XiRCON	$0.00
• File Compressor	Ultimate Zip	$0.00
• File Splitter	Chainsaw	$0.00
• FTP	Max-FTP	$0.00
• Help Editor	Microsoft HTML Help Workshop	$0.00
• Offline Browser	WinHTTrack	$0.00
• Searchbot	FirstStop WebSearch	$0.00
• E-mail Client, Newsreader, Telnet	Pegasus Mail, X-News, EasyTerm	$0.00
Total		**$0.00**

Source: Harrison Center Supply Closet <http://helios.acomp.usf.edu/~sharriso/supply-closet/index.html> and the writers/readers of the Usenet newsgroup alt.comp.freeware

human beings should be agents in the creation of their computer tools/environments; and, finally, that others of their own kind will join with them in common struggle to seize the tools to mark the world that marked them as other. Indeed, students in posthuman classrooms, when they have done with their work, will have perceived, however dimly, a political alternative to the present version of life in cyberspace. They will have participated in the counter-hegemonic manufacture of a cyborg self that is still a subject/object relation but that is not, at one and the same time, a commodity.

It would be too much to say that projects such as the one alluded to above will stand any chance of redirecting the trajectory of cyborg culture on its current path through a morass of self-acquisition, self-absorption and self-destruction. Yet, a pedagogy of the posthuman should awaken living commodities to the truth that capital has finally succeeded in turning life itself into a commodity fetish and, also, to the unlikely possibility that cyborgs will someday exist as something other than the manufacturers of the self that exists as slave to the self that lives in political isolation from selves who would free the self from the manufacturers of the self if only they were free.

NOTES

1. Those who MOO will no doubt be familiar with the terms e, em, eir, eirs, emself—namely, the "Spivak" neuter gender pronoun sequence. For readers new to these terms, the Spivak pronouns supersede the more common alternative—s/he, him/her, his/her, his/hers, and (him/her)self—and are useful to writers who have become uncertain of their capacity to attach dual sex/gender identities to socially constituted subjects. I decided to use Spivak pronouns when discussing my students because, given their increasingly intimate relationship with technology, they often seem as much like ambiguously gendered posthuman biotech workstations as they do biologic men and women.

2. *Omega* is derived from descriptions, pictures, and accounts taken from the Web site of MIT wearable-computing-specialist Steve Mann: http://www.eecg.toronto.edu/~mann.

3. Other kinds of cyborgs, principally those derived from pharmaceutically driven biotech applications, may appear in computer writing centers, overlapping in the bodies of computer-tech cyborgs, but these biotech cyborgs should have little impact, at least in this analysis, on the formation of posthuman *computer writing* pedagogies.

WORKS CITED

Baudrillard, Jean. "The System of Objects." *Selected Writings*. Ed. Mark Poster. Stanford, CA: Stanford UP, 1988.

Belcher, Diane D. "Authentic Interaction in a Virtual Classroom: Leveling the Playing Field in a Graduate Seminar." *Computers and Composition* 16 (1999): 253-67.

Blair, Kristine. "Literacy, Dialogue, and Difference in the 'Electronic Contact Zone.'" *Computers and Composition* 15 (1998): 317-29.

Buckley, Joanne. "The Invisible Audience and the *Disembodied* Voice: Online Teaching and the Loss of Body Image." *Computers and Composition* 14 (1997): 179-87.

Comstock, Michelle, and Joanne Addison. "Virtual Complexities: Exploring Literacy at the Intersections of Computer-Mediated Social Formations." *Computers and Composition* 14 (1997): 245-55.

Clynes, Manfred E., and Nathan S. Kline. "Cyborgs and Space." *The Cyborg Handbook*. Ed. Chris Hables Gray. New York: Routledge, 1995. 29-33.

Gabilondo, Joseba. "Postcolonial Cyborgs: Subjectivity in the Age of Cybernetic Reproduction." *The Cyborg Handbook*. Ed. Chris Hables Gray. New York: Routledge, 1995. 423-32.

Grabill, Jeffery T. "Utopic Visions, The Technopoor, and Public Access: Writing Technologies in a Community Literacy Program." *Computers and Composition* 15 (1998): 297-315.

Haefner, Joel. "The Politics of the Code." *Computers and Composition* 16 (1999): 325-39.

Haraway, Donna. "A Manifesto for Cyborgs: Science, Technology, and Socialist Feminism in the 1980s." *Socialist Review* 15.2 (1985): 65-107.

Harrison, Stanley D. "Cyborgs and Digital SoundWriting: Rearticulating Automated Speech Recognition Typing Programs." *Kairos* 5.1 (2000). http://english.ttu.edu/kairos/5.1/features/harrison/ (11 November 2000).

Johnson-Eilola, Johndan. "Control and the Cyborg: Writing and Being Written in Hypertext." *JAC Online* 13.2 (1993). http://jac.gsu.edu/jac/13.2/Articles/6.htm (1 April 2001).

———. *Nostaligic Angles: Rearticulating Hypertext Writing*. Norwood, NJ: Ablex, 1997.

Mann, Steve. "'Smart Clothing.'" 14 Feb. 1996. http://n1nlf-1.eecg.toronto.edu/smart_clothing/ (10 Oct. 2000).

Marx, Karl, and Frederick Engels. *The German Ideology: Part One*. Ed. C.J. Arthur. New York: International, 1970.

Selfe, Cynthia L. "Technology and Literacy: A Story about the Perils of Not Paying Attention." *College Composition and Communication* 50 (1999): 411–36.

Selfe, Cynthia L., and Richard J. Selfe, Jr. "The Politics of the Interface: Power and Its Exercise in Electronic Contact Zones." *College Composition and Communication* 45 (1994): 480–504.

Sullivan, Laura L. "Cyberbabes: (Self-)Representation of Women and the Virtual Male Gaze." *Computers and Composition* 14 (1997): 189–204.

Taylor, Todd. "The Persistence of Difference in Networked Classrooms: *Non-Negotiable Difference* and the African American Student Body." *Computers and Composition* 14 (1997): 169–78.

Whitaker, Elaine E, and Elaine N. Hill. "Virtual Voices in 'Letters Across Cultures': Listening for Race, Class, and Gender." *Computers and Composition* 15 (1998): 331–46.

PART TWO

Rhetoric and Writing in a Digital Age

4

THE AVAILABLE MEANS OF PERSUASION

MAPPING A THEORY AND PEDAGOGY OF MULTIMODAL PUBLIC RHETORIC

David M. Sheridan
Jim Ridolfo
Anthony J. Michel

> Many, perhaps most, of the university scientists who designed the architecture of the Internet did so with the explicit intent to create an open egalitarian communication environment. They had a vision of a noncommercial sharing community of scholars and, eventually, all citizens of the world. It was to be a public utility.
> —Robert McChesney

Whether or not the Internet will facilitate the emergence of a "sharing community" comprised of "all citizens of the world" remains to be seen. Equally uncertain are the limits to the possible forms that participation in this new community might take. What new types of discourses might citizens produce? What new roles might they assume? The answers to these questions are dependent, in part, on our conceptions of public rhetoric. As many have noted, the Internet and other digital technologies allow us to communicate not just through words, but also through sounds, colors, photographs, and other semiotic resources. What uses will public rhetoric find for these new affordances? What corresponding transformations will rhetorical education need to undergo?

Some of the possibilities and challenges that the Internet and other digital technologies offer public rhetoric are suggested by a short documentary film produced by Jim Ridolfo in response to the 2003 Free Trade Area of the Americas (FTAA) protests in Miami. In Jim's experience as an activist, mainstream media representations of protests were often characterized by a pattern of distortion resulting in part from the distanced perspective of the camera. Aerial shots from helicopters do not adequately capture the experience of facing police in riot gear. While Jim might have elected to send written critiques to mainstream media institutions, his experience with new media suggested a more independent and multimodal alternative. Wanting to place into circulation not just a comment on existing images but *new* images, Jim turned to video as a medium uniquely suited to making visible the experiences of the FTAA protestors. Using video footage that he had shot and digitized himself, Jim produced a short film and distributed it via the Internet. Within months, over 3,000 people had downloaded the film, and at least one independent bookstore began distributing it on CD, alongside more traditional leaflets and tracts.

Jim's FTAA digital video documentary can be partially understood in terms of a number of recent discussions within composition and rhetoric and related fields. The decision to adopt a multimodal approach reflects claims by various scholars that rhetorical education can be productively broadened to include semiotic resources other than the written word (see George; Kress and Van Leeuwen; Shipka; and Stroupe).[1] The decision to seek out an effective method of distribution is consistent with recent calls by other scholars for increased attention to the circulation (not just production) of compositions (see Trimbur; Shipka; Finnegan; and Welch, for example). The act of making the film public in the first place resonates with the field's re-awakened concern with public rhetoric (Weisser). The use of the Internet as a forum for addressing public issues can be understood in terms of recent work that explores the potential of the Internet to embody and extend our notion of the public sphere (see Ward and Hands). Finally, the decision to introduce into circulation images that counter those proliferated by dominant media resonates with a growing body of scholarship on the way marginalized groups appropriate multimodality to, as Anthony Ellertson says, "speak back to the popular culture surrounding them" (see, for instance, DeVoss; Goodman; Hawisher and Sullivan; Sheridan). This last group of scholars has been particularly helpful in theorizing the cultural work that can be accomplished when historically disenfranchised groups avail themselves of the opportunities afforded by new media.

Jim's deployment of multimodal public rhetoric lies, then, at the point where some of the most fertile discussions within the field of composition and rhetoric converge. And yet, both the civic and the multimodal continue to be integrated into our classrooms in reductive, limiting ways. As Douglas Hesse observes, "We often have tended to distill civic writing into a school genre. That is, we have students write *about* the civic sphere, not *in* it. In like fashion, our new [. . .] fondness for visual rhetoric manifests itself considerably more in the analysis of, rather than in the production of images" (350). We submit that if multimodal public rhetoric has not yet been fully and richly integrated into our pedagogies and curricula, it is because we do not yet have a coherent plan for confronting "semiotic resources" as, in the words of Maureen Goggin, "multiple complexes of technological conditions and sociocultural landscapes that overlap like Venn diagrams" (89). Nor do we have a plan for confronting key shifts in the way rhetorical labor is divided—shifts that result in a new kind of rhetor who not only speaks and writes, but also designs and publishes.

In our admittedly utopian vision, the public sphere becomes a space where nonspecialists self-reflexively engage in an extended "conversation" characterized by the rhetorically effective integration of words, images, sounds, and other semiotic elements—a space, in other words, where rhetorical interventions like Jim's protest film are commonplace. What is needed, we contend, is a conceptual apparatus—a map—that better enables rhetorical educators to confront a host of fundamental and concurrent shifts in rhetoric as a simultaneously symbolic, cultural, and material practice that occurs within local and extended contexts that are themselves simultaneously discursive, cultural, and material.

Our map begins with the complex issue of access, which can be productively explored by looking at the case of an old media technology: the still camera. Our discussion of access as a material and cultural issue ultimately leads us to suggest several key points of critical engagement that rhetorical education might take up, including a critique of writing as a privileged mode. We find in the ancient rhetorical concept of kairos—the "opportune moment"—a theoretically coherent entry point for expanding a rhetor's assessment of "the available means of persuasion" to include assessments of the material and discursive conditions that shape the production, distribution, and reception of the rhetor's argument. In Jim's case, for instance, a successful rhetorical intervention required him to assess the affordances of digital video relative to other media and modes (a written essay, a tri-fold brochure, a website) in terms of both its appropriateness for his intended audience and the resources to which he had access.

In exploring this more expansive response to the "opportune moment" in the age of new media, we find ourselves compelled to remap one of the key tensions embedded within the rhetorical tradition: the tension between a deliberative rhetoric—characterized variously as civil, rational, and conciliatory—and a confrontational or agonistic rhetoric. Addressing the way multimodality implicates deliberative rhetoric leads us to current dialogues in visual ethics, visual anthropology, and critical semiotics. As a case in point, we focus on the photograph and its tendency to resist inspections of its own rhetoricity—a tendency, we argue, that is at odds with the goals of deliberative rhetoric. We explore the possibility that critical self-reflexiveness provides a way of recovering deliberative goals for multimodal rhetoric.

Wanting to preserve for public rhetors the ability to make strategic choices in response to a variety of rhetorical situations, we do not foreclose the possibility of deploying a more confrontational multimodal rhetoric. Indeed, post-Habermasian models of the public sphere (Chantal Mouffe's, for example) reveal the need for confrontation. We see Umberto Eco's trope of "semiotic guerilla warfare" as suggestive of an important kind of cultural work that multimodal rhetoric can perform—a kind of cultural work recently acknowledged by public-sphere theorists such as Scott Welsh and Kevin DeLuca. Welsh observes that "an effective democratic challenge must be geared to affect and effect prevailing cultural vocabularies" (685). Jim's deployment of filmic rhetoric can be understood as this kind of challenge: the visual vocabulary and syntax deployed by the mass media—often characterized by "objective" images of demonstrations captured from the distanced perspective of a helicopter—are countered in his film with eye-level footage taken from the perspective of the demonstrators themselves. This is consistent with DeLuca's observation that the "unorthodox rhetoric" of what he calls "image events" "reconstitutes the identity of the dominant culture by challenging and transforming mainstream society's key discourses and ideographs" (16).

Finally, we address the institutional and disciplinary constraints that rhetorical education must engage if it is to take up the project of multimodal public rhetoric. We explore the possibility that by taking advantage of what Rolf Norgaard calls "disciplinary contact zones," rhetorical education can foster a praxis-based approach to interdisciplinarity. Embracing such contact zones, however, means substantially reconfiguring key institutional structures and practices as well as attitudes toward disciplinarity itself.

In contradistinction to skeptics like Neil Postman, John Phelan, and others who fault visual media for their tendency to "erode political space and reduce participation in democratic processes" and to "numb our minds," Bruce McComiskey asks us to imagine a new public sphere that exploits the "countless positive and, indeed, liberating uses of images in communication" ("Viusual" 193–94, 200). The following discussion is an attempt to sketch out what this public sphere might look like and what reconfigurations of rhetorical education might make it possible.

RETHINKING ACCESS AND THE PUBLIC SPHERE

The kind of multimodal public sphere we imagine is contingent upon nonspecialist citizens having access to an array of cultural and material resources, including technologies, knowledge bases, and skill sets. To understand the complex issue of access at the present moment, it is useful to examine analogous cases in history, such as the case of the still camera, made widely accessible to nonspecialists in the late nineteenth century when George Eastman invented "snapshot" photography by combining roll film with his small, portable "Kodak." The case of the still camera provides important insights into how cultural and material logics circumscribe the adoption of emergent technologies and rhetorics.

In his 1909 talk "Social Photography, How the Camera May Help in the Social Uplift," reform photographer Lewis Hine implores his audience to use photography as a political tool. Iconic rhetoric, Hine claims, "brings one immediately into close touch with reality," and the photograph in particular "has an added realism of its own," an "inherent attraction not found in other forms of illustration" (111). Hine remarks that although his own era belongs to the "specialist," there is much to be gained "by the popularizing of camera work" (112).

Contemporary theorists such as John Tagg and Don Slater have acknowledged the potential power of the camera, as a technology available to nonspecialists, to effect social change. In Slater's words,

> The camera as an *active* mass tool of representation is a vehicle for documenting one's conditions (of living, working and sociality); for creating alternative representations of oneself and one's sex, class, age-group, race, etc.; of gaining power [. . .] over one's image; of presenting arguments and demands; of stimulating action [. . .]. (290)

Slater and Tagg, however, agree that a tradition of radical mass photography has not developed on a large scale. Tagg explains that relevant technologies "only passed into popular hands in the crudest sense of the term" (17). Amateurs have remained dependent on large corporations for many aspects of the photographic process. Additionally, important "technical and cultural knowledge" continues to reside in the hands of professionals and corporations (18). Finally, amateur photography has been situated within a cultural hierarchy that privileges professionals and artists while it relegates amateurs to the lower registers of "kitsch" (19). Slater adds that consumers have been conditioned through "high pressure mass marketing of photographic equipment" that restricts the use of cameras to nonpolitical purposes (290). The "Kodak moment" is not a moment of civic intervention but of individual sentimentality. For all of these reasons, nonspecialists have tended to see cameras as technologies of leisure rather than technologies of civic engagement. The "enormous productive power" of the camera "is effectively contained as a conventionalized, passive, privatized and harmless leisure activity" (Slater 289). The case of the camera suggests that access to the public sphere is not a simple issue of making material resources available, but involves a complex set of relationships between material, cultural, and economic factors.

The issue of access can be more fully understood by locating it within ongoing conversations about the nature of the public sphere. In popular discourse "public sphere" sometimes refers in a general way to mass media and mass culture. In this sense, of course, the public sphere has been characterized by multimodal rhetoric for some time, dominated as it is by television, film, and radio. This is not, however, what we mean by "public sphere." Following an established—if conflicted—line of social theorists that include Hannah Arendt, Jürgen Habermas, Seyla Benhabib, and Nancy Fraser, we mean, instead, a rhetorical space in which "the citizens of a pluralistic polity speak from and across their differences productively" (Ivie 278).

Key to the public sphere in this more limited sense is the issue of inclusiveness. Even Habermas, often criticized for his narrow focus on white affluent men, prescribes what Benhabib calls a "symmetry condition," which includes the related tenets that "each participant must have an equal chance to initiate and to continue communication," and "each must have an equal chance to make assertions, recommendations, explanations, and to challenge justifications" (87). Likewise, Craig Calhoun writes that "[i]n a nutshell, a public sphere adequate to a democratic polity depends upon both quality of discourse and quantity of participation" (2). While mass media like television, radio, and film might be public in the sense of addressing a wide audience, they have not historically offered symmetrical opportunities to participate. Until recently, most citizens have not had access to the means of production and distribution associated with film, television, or radio, nor has the primary purpose of these media been to extend a conversation among concerned citizens (see Habermas 183–85).

The issue of media access has become even more complex with the emergence of digital technologies. In her recent interrogation of the so-called "digital divide," Barbara Monroe argues that asymmetrical conditions of access to technology need to be understood in the context of broader cultural and material realities. Reminding us that class inequality "is at once economic, racial, discursive, and epistemological in character," Monroe suggests that "[r]esituating the [digital] divide within the landscape of larger social and political formations should allow for a richer, more complicated discussion of a host of issues that attach themselves to Internet access per se but are actually constituted by these larger formations" (5). Monroe's analysis of access to digital technologies echoes the analysis of the still camera by Tagg and Slater in its emphasis on the need to move beyond material conditions to larger cultural structures. Monroe concludes that a pedagogy of critical engagement is both necessary and possible (29-30).

Before turning to the issue of critical engagement, however, we would like to explore more fully the complex and shifting nature of material access. For if material access is not a sufficient precondition for the development a multimodal public sphere, it is certainly a necessary one. As Nicholas Garnham observes, contemporary models of the public sphere must include provisions for "the problem raised by all forms of mediated communication, namely, how are the material resources necessary for that communication made available and to whom?" (361). Even in this more narrow sense, access is not a simple matter. Ready access to handheld cameras, as Tagg points out, belied a lack of access to other key technologies related to photographic communication. Likewise, Monroe reminds us that there is no such thing as generic access to computers: access at home is not the same as access at work or at school (19-20, 26-27).

Historically, mediated communication—and the multimodal rhetoric associated with it—has been a one-way street, relegating nonspecialists to the role of consumers. If technologies associated with the still camera have historically been withheld from nonspecialists, the technology gap has been even more pronounced in the case of film, television, radio, and print. The expensive and arcane technologies of media production have required professionalized and elaborately divided labor. Producing content for these media has historically necessitated an ensemble of specialists (graphic designers, photographers, cinematographers, script writers, producers, directors, editors) funded by large corporations. Additionally, technologies of distribution have been designed to facilitate communication of the few to the many rather than from the many to the many. Echoing Bertolt Brecht's observations about radio, Hans Enzensberger contends that "[i]n its present form, equipment like television or film does not serve communication but prevents it. It allows no reciprocal action between transmitter and receiver" (97).

As writers such as Richard Lanham have noted, however, emergent technologies are altering this media asymmetry by providing nonspecialists the resources necessary to create rhetorically effective multimodal compositions. The personal computer allows nonspecialists to manipulate visual and aural semiotic elements in ways historically reserved for highly trained specialists. Communicators who hope to make use of photographs, for instance, can turn to a host of free or inexpensive applications that allow them to crop and zoom, adjust color saturation, lightness, opacity, contrast, sharpness and so on. Likewise, many of the editing operations crucial to rhetorically effective uses of film—the ability to sequence footage, to cut between shots, to add music—are now easy to perform using a standard computer and free or inexpensive software. Nonspecialists increasingly have access to applications that allow us to draw, paint, compose music, and create animations.

Just as important as technologies of production are technologies of *reproduction and distribution*. Historically, individuals and small groups who have managed to secure the resources to produce television content, for instance, have been left with the problem of how to deliver that content to target audiences. Airtime is costly, and even those who can afford it are usually limited to short "spots" that run only a few times. Likewise, resources available to corporate print media (color inks, photographs, bindings) add exponentially to the cost of reproduction. Problems related to reproduction and distribution, however, are increasingly addressed by the Internet and other digital technologies. Communicators can distribute a wide range of multimodal content via the Internet for a tiny fraction of what it would have cost in the past. Colors, images, and other semiotic elements do not add to the cost of reproducing digital compositions. A standard personal computer can distribute millions of copies of a web page without incurring additional cost. In contrast to television, film, or radio, content on the Internet can be made available twenty-four hours a day without adding costs and without displacing other content. Further, the Internet is a many-to-many, not one-to-many, technology. The predominant metaphor for new media is not a pipeline distributing content from a central location to dispersed individuals, but a web or network in which media elements are joined through hyperlinks.

In short, nonspecialists, including historically disenfranchised groups, can potentially own or access the means of production, reproduction, and distribution of media content on a much larger scale than in the past. The division of media labor, as John Trimbur notes, is "collapsing" ("Delivering" 269); the same individual can produce and distribute multimodal compositions to a mass audience. It is as if when George Eastman began putting Kodak cameras into the hands of consumers in 1888, he also provided them with easy-to-use darkrooms and printing presses as well as access to an elaborate system of delivery trucks and retail outlets.

But, of course, access is not just a function of material resources. As both Tagg and Monroe demonstrate, citizens will not be able to exploit available material resources unless they have access to related knowledge and skill sets. As we discuss later, fostering the kind of knowledge and skills necessary for nonspecialist appropriation of multimodal rhetoric requires a critical praxis of interdisciplinarity. Even more fundamental, however, are several new points of critical engagement that rhetorical education (among other disciplines) might integrate. First, rhetorical education needs to open up for critique the *inevitability and privilege of written rhetoric and associated media and technologies* within the academy.[2] The historical insistence of official education in general and rhetorical education in particular on the written word as the only legitimate rhetorical practice habituates citizens to a kind of learned helplessness; sustained insistence on the written word renders invisible to citizens the possibility of deploying multimodal rhetoric and, therefore, withholds from them a potentially transformative set of rhetorical assets.

Secondly, citizens need to critique the logics of profit and consumerism that enforce an overly narrow understanding of multimodality and technologies associated with it. As Tagg, Slater, and Monroe demonstrate, the logic of capitalism circumscribes our perceptions of how modes, media, and technologies can be used. Like the still camera, digital technologies continue to be constructed as tools for professional and personal—not civic—spheres. Computers and the WWW are often marketed as commodities that allow us to work efficiently, play games, chat with family and friends, and engage in more consumption.[3] Rhetorical education needs to provide opportunities for public rhetors to re-imagine multimodal rhetoric as a tool for addressing public concerns.

Finally, a critically reflective approach needs to open up intellectual spaces for students to critique the *division of labor* in rhetorical production. In the case of the still camera, most nonspecialists never questioned the practice of taking rolls of film to the local one-hour photo processor. The separation of the labor of shooting a picture from the labor of processing it has become naturalized. But as Tagg's analysis shows, this division curtailed the development of radical photography because it removed from nonspecialists the tools necessary for effective visual communication. Speaking more generally, the division of labor in rhetorical production is often viewed as a given, when in fact it is a function of cultural, material, and historical forces. It used to be the case, in certain contexts, that communicators handwrote or dictated compositions; someone other than the writer (often someone in a subordinate role) was responsible for the labor of typing. Word processing technologies have rendered this practice largely obsolete. In the case of mediated communication, this division of labor has been elaborate, involving highly trained specialists. But the material conditions that necessitated this have shifted in important ways, and this division of labor needs to be fundamentally reimagined.

In the transformation of rhetorical education that we envision, these fundamental dynamics—the academy's privileging of the written word; the cultural logics that circumscribe the use of certain modes, media, and technologies; and the division of rhetorical labor—would be exposed for scrutiny. These dynamics would be reconceived in terms of the needs and goals of public rhetorics. In the following section we begin this reconception by turning to the ancient concept of kairos, which offers a coherent way of integrating decisions about mode, medium, and associated technologies into rhetorical theory and practice.

MODE, KAIROS, AND MATERIALITY

> In the age of digitization, the different modes have technically become the same at some level of representation, and they can be operated by one multi-skilled person, using one interface [...] so that he or she can ask, at every point: "Shall I express this with sound or music," "Shall I say this visually or verbally?", and so on.
> —Gunther Kress and Theo Van Leeuwen

The points of critical engagement that we outline above have in common the goal of expanding the rhetorical agency of public rhetors, of making visible options that have historically been foreclosed by various cultural-material logics. The concept of kairos—defined by James Kinneavy as "the appropriateness of the discourse to the particular circumstances of the time, place, speaker, and audience involved," offers a coherent way for rhetors to assess these new options (84). By *materializing* kairos, we can render visible a variety of options that have historically been elided. That is, kairos can be productively expanded to include assessments of modes, media, and the technologies of production, reproduction, and distribution associated with them.

In order to understand this expansion, we need to bear in mind the essentially performative nature of rhetoric. Rhetoric, Lloyd Bitzer famously reminds us, "functions ultimately to produce action or change in the world; it performs some task. In short, rhetoric is a mode of altering reality [...]" (3–4). In the case of civic rhetoric, an intervention is invited by an exigency of public concern. Some social condition—poverty, intolerance, inequality—needs to be changed and rhetoric is a means of accomplishing this change.

Key sites of rhetorical education within the academy, such as first-year composition, have often begun with the hegemonic assumption that paper-based alphabetic rhetoric was the appropriate—or even the only possible—response. But public rhetors cannot afford to begin with that assumption. A grass-roots organization, in approaching a particular social exigency, does not begin with the question "What kind of paper should we write?" Instead, it assesses the available means of persuasion much more broadly and strategically in terms of the materials necessary for the production and distribution of appropriate rhetorical compositions within a particular set of circumstances: What resources are available to us at this moment? What modes and media are both within our means and are best suited to our audience and purpose? Nor can a grass-roots organization afford the luxury of focusing narrowly on composing. A composition, in itself, does not address an exigency; to effect change, it needs to be delivered to its intended audience. Organizations, therefore, must ask, What medium of distribution (direct mail, television, radio, and so on) will allow us to get our message to our audience most effectively?

Jim Ridolfo's decision to employ digital video points to the utility of making kairotic assessments related to the production of multimodal compositions. Confronted with a particular exigency (the potential harm caused by representations of protests produced and distributed by the mass media), Jim had to decide what form of rhetorical intervention would be most effective. This intervention might conceivably have been a white paper, a website, a brochure, a poster, or a leaflet. Digital video, however, was suited to his particular subject and purpose because he sought to participate in discourses—mainstream media representations—that employed visual and aural elements. To frame the mass media representations *as* representations, Jim opted for the strategy of providing alternative representations aimed at making visible a different perspective, video footage that would allow him to partially capture the visceral experience of demonstrating. The exigency itself, in this case, related to particular practices of video mediation. Digital video was also appropriate for Jim's intended audience, a diverse (mostly young) group of alternate globalization activists spread across the hemisphere.

If rhetoric's aim, as Aristotle says, is discovering the best available means of persuasion, processes of production need to be expanded to include assessments of semiotic resources (modes) as well as material resources (technologies, raw materials, media, time). Assignments that begin with the directive "write a paper" curtail the rhetorical and ultimately the civic agency of student-citizens because they elide an important set of decisions that public rhetors must face in a digital age. Rather than allowing students to assess modes, media, and the material constraints associated with them, framing assignments in this way imposes on students the mode (writing), the medium (paper), and the technologies involved (pens, pencils, word processing applications).

Here we are following Jody Shipka's recent attempt to create a framework "geared toward increasing students' rhetorical, material, and methodological flexibility" (285–86). As Shipka observes, "assignments that predetermine goals and narrowly limit the materials, methodologies, and technologies that students employ in service of those goals [. . .] perpetuate arhetorical, mechanical, one-sided views of production" (285). Our emphasis, however, differs from Shipka's in important ways that reflect our aim of connecting multimodal rhetoric with public-sphere participation. Despite Shipka's emphasis on a "task-based" approach, her framework does not foreground assessments of mode that are informed by exigency, purpose, and audience. Shipka adopts an essentially expressivist approach in which students select modes and media in order to communicate their feelings and experiences. For instance, assigned the task of creating a composition based on the *OED*, one of Shipka's students experiences "physical and intellectual punishment [. . .] while sitting in front of the computer looking for usable OED data online" (297). To express this frustration, he produces a video tailored to "bore the socks off" his audience (297). Likewise, another student, confessing "that she was extremely frustrated for the first part of the semester," sees the *OED* assignment as "her opportunity to articulate that frustration through a piece that was intentionally designed" to instill frustration in her audience. The framework that Shipka offers is limited by her expressivist approach. Public rhetoric depends on the rhetor's ability to move beyond herself, to negotiate through a complex set of relationships between audiences, purposes, material circumstances, and exigencies. In short, public rhetoric demands a kairotic approach.

In addition to concerns of production, kairotic assessments of modes and media—like assessments of when to appeal to pathos, ethos, or logos—require that public rhetors confront issues of both the reproduction and distribution of rhetorical compositions. As he considered the possibility of producing a film, for instance, Jim needed to assess the complex relationship between production and distribution. In a pre-Internet era, video might not have been a suitable medium for Jim's response, not because video was inappropriate for a particular audience and purpose, but because it was impractical to distribute. The affordances of digital networks, which make digital video relatively easy to distribute, encouraged the use of this medium. Moreover, the specific mechanism of delivery—the file-sharing application LimeWire—supported his rhetorical goals. LimeWire's peer-to-peer file-sharing approach allows for more precise searches by those seeking a particular media format. Peer-to-peer file sharing allows direct access to content in ways that the web itself does not; media elements are not embedded within larger structures as they are on webpages, but are directly accessible through a search restricted to keyword and media format. Finally, Jim knew that if he made his footage available in a high-quality format, other media producers would be able to appropriate his footage for their own projects—a prospect that itself could potentially further his activist-minded rhetorical goals. Before composing, Jim had to anticipate the ways in which his film might be appropriated and to decide whether or not and how to facilitate these processes of sub-composing.

Opportunities for making decisions about reproduction and distribution in the traditional writing classroom are even more limited than opportunities for assessing modes and media.[4] As John Trimbur observes, "By privileging composing as the main site of instruction, the teaching of writing has taken up what Karl Marx calls a 'one-sided' view of production, and thereby has largely erased the cycle that links the production, distribution, exchange, and consumption of writing" ("Composition" 190). Indeed, the shift in focus from speech to writing that has defined contemporary rhetorical education for most citizens has made the elision of production and distribution even more pronounced than in classical times. Mindful that the effectiveness of a piece of oratory was dependent upon an embodied performance—including such elements as tone, volume, facial expression, and gesture—ancient rhetoricians studied the canon of delivery. In the traditional writing classroom, as Trimbur points out, delivery is reduced to "an afterthought at best" (190). "Reproduction" typically means printing out a paper and "distribution" means handing it to the teacher. In contrast to this reduction, Trimbur links delivery to public-sphere participation, arguing that delivery must be seen as "ethical and political—a democratic aspiration to devise delivery systems that circulate ideas, information, opinions, and knowledge and thereby expand the public forums in which people can deliberate on the issues of the day" (190).

Considerations of production and distribution and the technologies associated with them point to the material nature of rhetoric, a concern to which rhetoric and adjacent fields have recently turned their attention (see, for instance, Ellertson and Graham; Kress and Van Leeuwen; Selzer and Crowley; and Wysocki). This recent interest in materiality can be seen as a corrective to rhetorical studies' historically narrow focus on rhetoric's "most ephemeral quality: symbolicity" (Blair 18). Our goal here is to foreground the implications of materiality for multimodal rhetoric as a civic tool. This means preparing citizens to confront materiality as *producers* of rhetoric. McComiskey alludes to this concern in his review of *Rhetorical Bodies*, lamenting that "there is little attention paid" in the collection

> to the actual composition of material rhetoric. [. . .] Rhetoric and composition is primarily a productive art, an art aimed at the invention, arrangement, and delivery of cultural meaning—whether through voice, text, or image. Critical knowledge is "good" only insofar as it is "useful," and it is useful only insofar as it leads to positive rhetorical interventions into the material and discursive processes of oppressive political formations. (703)

It might have been possible for rhetorical education to overlook materiality in a writing-centered culture in which multimodal rhetoric was beyond the reach of nonspecialists and the tasks of distribution belonged to individuals other than composers. But our options and resources in a digital age have broadened significantly (see Kress and Van Leeuwen 2; Lanham; Wysocki 10). Educational sites designed to prepare students to produce multimodal public rhetoric need to help students negotiate the material processes associated with effective rhetorical intervention.

To sum up, in order to make kairotic assessments of mode, media, and the technologies of production, reproduction, and distribution associated with them, students need to confront such questions as, What modes and media are best suited to the kinds of change I am trying to effect and to my intended audience and purpose? How can I deliver my proposed rhetorical composition to my target audience? What material resources—technologies, knowledge, people, time—are available to me? What material limitations and affordances must I take into account?

OPEN DELIBERATIVE RHETORIC AND MULTIMODALITY

We have been arguing that rhetorical education can help prepare nonspecialists to deploy multimodal rhetoric effectively within the public sphere by adopting a more capacious understanding of kairos to include material considerations related to modes, media, technologies, and other resources. We do not mean to imply, however, that claiming multimodality for rhetoric is a matter of simple addition, as if public rhetors merely need to make a few extra decisions about images, computers and networks over and above traditional decisions about rhetorical strategy. Indeed, the material and cultural specificities of multimodal rhetoric force us to reinterrogate ongoing debates surrounding the nature of public rhetoric itself. One of the chief tensions within these debates is between models that emphasize a more deliberative approach (characterized variously as civil, rational, and conciliatory) and those that emphasize a more confrontational or "rowdy" one (Ivie 277). Arguing for the former are theorists as diverse as Deborah Tannen, Jürgen Habermas, Iris Marion Young, Richard Fulkerson, and Edward Corbett, while those who reserve a place for a more confrontational approach to public debate include Susan Jarratt, Gerald Graff, Kevin DeLuca, Ernesto Laclau, Chantal Mouffe, and even Ken Macrorie.

Lynch, George, and Cooper argue for an inclusive model of rhetoric that balances conciliatory and combative moments: "What we are seeking is a way of reconceiving argument that includes both confrontational and cooperative perspectives, a multifaceted process that includes moments of conflict and agonistic positioning as well as moments of understanding and communication" (63). Likewise, we hope to preserve for public rhetors the possibility of kairotically assessing their position on a continuum between deliberative and confrontational approaches. In this section we explore the way multimodality implicates deliberative rhetoric, turning in the following section to what Umberto Eco has called "semiotic guerilla warfare" (qtd. in Hebdige 105).

Evoking a tradition of deliberation, Corbett famously advocates an "open-handed" rhetoric characterized by "the kind of persuasive discourse that seeks to carry its point by reasoned, sustained, conciliatory discussion of the issues" in contrast to closed-fisted rhetoric which "seeks to carry its point by non-rational, nonsequential, often nonverbal, frequently provocative means" (288). Recalling Corbett's open hand, Richard Fulkerson notes that argumentation is the "chief cognitive activity" by which a democracy functions, and he attempts to outline a pedagogy in which argument is seen

> in a larger, less militant [. . .] context—one in which the goal is not victory but a good decision, one in which all arguers are at risk of needing to alter their views, one in which a participant takes seriously and fairly the views different from his or her own [. . .] . It is crucial that students learn to participate effectively in argumentation as a cooperative, dialectical exchange and a search for mutually acceptable (and contingent) answers. (16)

Fulkerson and Corbett point to a rhetorical tradition in which deliberators are open to change, to opposing views, to the critical scrutiny of others, in contrast to an agonistic tradition in which opponents seek to defeat each other, even if that means deploying rhetorical strategies that tend to resist, rather than invite critical scrutiny.

In order to illustrate the way multimodality changes the dynamics of deliberation, we return to our discussion of the photograph. In the tradition of critical semiotics stemming from Roland Barthes, photographs and other forms of iconic rhetoric are understood to be especially deceptive in their tendency to masquerade as transparent and authoritative representations of reality. Photography theorist Victor Burgin alludes to this tendency in a famous passage:

> More than any other textual system, the photograph presents itself as "an offer you can't refuse." The characteristics of the photo-

> graphic apparatus position the subject in such a way that the object photographed serves to conceal the textuality of the photograph itself—substituting passive receptivity for active (critical) *reading*. (146)

Rather than reveal itself as a rhetorical artifact, a photograph tends to assert itself as a given, a transparent window on reality, as natural, objective, neutral, authoritative, true, and real. Virtually every inquiry into how photographs mean is forced to confront this tendency at some point. Barthes' own observation is that "the denoted image naturalizes the symbolic message, it innocents the semantic artifice of connotation [. . .]" (45).

As Paul Messaris points out, a photograph's pretense of objectivity stems from its status as both an "index" and an "icon" (xii–xvii). As an index, in the Piercean sense, a photograph derives authority from the fact that it seems to be caused by the thing it represents. Light reflects off an object, enters the lens, and causes a photochemical reaction with the film that results in what Susan Sontag calls a "trace" of the real (154). As an icon, a photograph derives authority because our process of decoding it seems to be coequal with our process of decoding the world itself. Even Umberto Eco, who denies that a photograph is in any way an "analogue" of reality, concedes that "we perceive the image as a message referred to a given code, but this is the normal perceptive code which presides over our every act of cognition" (33, 32).

Film—especially documentary film—shares the iconicity and indexicality of the still photograph. Visual Anthropologist Jay Ruby observes that

> the filmic illusion of reality is an extremely dangerous one, for it gives the people who control the image industry too much power. The majority of Americans [. . .] receive information about the outside world from the images produced by film, television, and photography. If the lie that pictures always tell the truth is perpetuated [. . .], then an industry that has the potential to symbolically recreate the world in its own image continues to wield far too much power. (149)

Photographs and film as media and modes are not neutral; some rhetorical goals are more easily achieved than others as a result of the material realities of iconic media and the way they are constructed within Western cultures. Whereas deliberative rhetoric reveals its strategies in order to facilitate cooperative dialogue, photographs and films conceal their rhetorical nature, pretend, in fact, not to be rhetorical objects at all. Whereas open deliberative rhetoric reveals the status of any authorities it cites so that participants in the dialogue can critically assess them, photographs and film tend to assert themselves as authoritative, as mechanical reproductions of reality that cannot lie. As "offers you can't refuse," the photograph and film are the opposite of cooperative rhetoric.

But photographic authority is a tendency, not an inevitability. Michael Shapiro observes that "photography plays a politically radical role when it opens up forms of questions about power and authority which are closed or silenced within the most frequently circulated and authoritative discursive practices" (130).

Self-reflexivity is one strategy for achieving this alternate possibility in which photographs *denaturalize* reality, revealing rather than masking hidden assumptions. "To be reflexive, in terms of a work of anthropology," Jay Ruby explains, "is to insist that anthropologists systematically and rigorously reveal their methods and themselves as the instrument of data generation and reflect upon how the medium through which they transmit their work predisposes readers/viewers to construct the meaning of the work in certain ways" (152). Reflexivity, for Ruby, is essential: "the maker of images has the moral obligation to reveal the covert—never to appear to produce an objective mirror by which the world can see its 'true' image" (140). Ruby outlines a variety of rhetorical strategies for achieving reflexivity, including filmic techniques that remind audiences of the camera's operation. Film makers can eschew the "voice of god" narrator and instead foreground uncertainty (for example, by including discussions in which those involved in the production of the film argue about what has happened). They can include multiple retellings of a single event so that audiences realize that meaning is dependent upon the way an event is depicted (115–35).[5]

The website *360degrees: Perspectives on the Criminal Justice Systems* illustrates the way multimodal compositions can adopt a self-reflexive approach. The self-announced purpose of this site is to "challenge your perceptions about who is in prison today and why" (Cornyn et al.). The title of the site, alluding to a circle, foregrounds the importance of taking a holistic view, one that encompasses a variety of perspectives. The trope of the circle is visually reinforced by a number of design elements. The site's menu, for instance, is a series of floating circles labeled "Timeline," "Stories," "Dialogue," and so on. The section of the site labeled "Stories" includes seven cases of incarceration. These cases are not represented by a single authoritative voice, but by a collection of voices, including the individual who was incarcerated, representatives of the criminal justice system (judges, prosecutors, guards, juvenile program workers), victims, and relatives of victims and convicts. These individuals are represented visually by still photos. Each photo (itself framed as a circle) is positioned around a larger circle, providing an image of round-table dialogue. Clicking on one of the circles calls up an audio recording of a monologue in which the individual voices his or her understanding of the case. These audio recordings are accompanied by a video in which a camera films a 360 degree circuit of a physical space relevant to the person talking (a prison cell, an office, a bedroom).

The *360degrees* website uses the figure of the circle to foreground the idea of perspective, to communicate that these issues and cases cannot be reduced to a single authoritative view. Moreover, by adopting a prominent metaphor that is iterated in words, design elements, and video, the website foregrounds the act of representation itself. By making its key metaphors explicit and by continually drawing attention to them, the website points to its own rhetoricity.

Rhetorical education can facilitate the project of a multimodal public sphere by making citizen communicators aware of the naturalizing tendencies associated with certain kinds of multimodal rhetoric and can prepare them to use strategies like reflexivity in order to achieve the goals of public deliberation. Confronting the ethical implications of iconic rhetoric, however, is only one example of the new challenges rhetorical education faces as it confronts multimodality. Emotion, for instance, is differently operationalized by multimodal rhetoric (Hill 30–38), calling for a better understanding of the roles that pathos plays in deliberation. And technical communication theorists like Sam Dragga, Nancy Allen, and Donna Kienzler demonstrate that rhetorical choices related to the visual presentation of information have ethical implications.

We imagine an approach to rhetorical education in which the public sphere is characterized by nonspecialists making strategic decisions about the kind of multimodal rhetoric they will deploy. In situations that call for rhetors to build consensus and facilitate cooperation, they will need to confront the way multimodality changes the nature of deliberation. They will need to ask questions like, How do the particular meaning-making processes associated with this mode and medium aid or resist critical reflection? What is naturalized by this particular confluence of semiotic elements? How can I integrate semiotic elements in such a way that their rhetorical nature is foregrounded rather than hidden?

SEMIOTIC GUERILLA WARFARE: THE CULTURAL WORK OF MULTIMODALITY

> Ethnic minorities, women, gays, third- and fourth-world people, the very rich, and the very poor are telling the middle-class, middle-aged straight white males who dominate the industry that the mass-mediated pictures of the Other are false. Many wish to control [. . .] the ways in which they are imaged by others. —Jay Ruby

> The potential power that you possess as a writer is based on this ability to put images of the world *you* see, and ideas about these images, into other people's heads.
> —Wayne Booth and Marshall Gregory

Like Lynch, George, and Cooper, we hope to preserve an understanding of public rhetoric that "includes both confrontational and cooperative perspectives, a multifaceted process that includes moments of conflict and agonistic positioning as well as moments of understanding and communication" (63). In keeping with the kairotic approach we outline, we argue that to participate in effective rhetorical interventions, public rhetors should strategically position themselves in relation to their audience and subject matter based on assessments of contextual factors such as exigency and purpose. Although Habermas and others have emphasized the "rational" and the "civil" as governing tropes for public rhetoric, other scholars have pointed out the need for confrontation. In this section we borrow Umberto Eco's figure of "semiotic guerilla warfare" to theorize the cultural work that multimodal rhetoric can perform when a more confrontational approach is adopted.

Our exploration of confrontational multimodal rhetoric begins with a particular understanding of the relationship between multimodal signifying practices and consciousness. Cultural theorists from Edward Sapir to Michele Foucault have emphasized the ways in which discourse, broadly conceived, is productive of thought, ideology, and culture itself. Hans Enzensberger observes that while all of us "like to think that we reign supreme in our own consciousness, that we are masters of what our minds accept or reject," our consciousness is actually the product of a "consciousness industry" in which mass media play a privileged role (3, 95).

Jennifer González, for instance, helps us understand the way racial encoding is accomplished by the consciousness industry's deployment of particular modes and media. In "Morphologies: Race as a Visual Technology," González writes, "Skin color, hair color, and eye color become marking devices for those who seek to situate the genetic history of humans within the narrow confines of phenotype. Race has always been a profoundly visual rhetoric [. . .]" (380). According to González, photographic technologies have been particularly implicated because "a conceptual parallel exists between the 'truth effects' of photography and what might be called the 'truth effects' of race. Both kinds of 'truth effects' naturalize ideological systems by making them visible and, apparently, self-evident" (379–80).

Rhetorical education and the humanities more generally have historically emphasized critique as the appropriate response to the practices of the consciousness industry. For instance, McComiskey draws on cultural studies to provide students a heuristic for uncovering the values embedded in both visual and verbal semiotic resources deployed by magazine advertisements. Importantly, however, McComiskey asks his students to go beyond what might be called academic critique—the production of critical discourse in the classroom—to critique as rhetorical action in the world beyond the academy. Mindful that social change is "less likely to occur if students end their composing processes with critical essays," McComiskey has students make a "rhetorical intervention" in the form of a "practical letter" targeted at one or more relevant audiences: the company that commissioned the ad, the editors of the magazine that distributes it, or the target audience for the ad (391). In selecting their audience, the students assess the "potential impact" of their rhetorical intervention (394).

McComiskey's emphasis on written rhetorical intervention can be usefully synthesized with his later emphasis on students as producers of multimodal discourse. If, as McComiskey notes, Calvin Klein ads promote the restrictive ideal of "sexy women wear[ing] close fitting jeans," one powerful rhetorical response would be the creation and distribution of alternative images (393). The women in the class might, for instance, produce images of womanhood that are aligned with their own critical understandings and lived experiences and might publish these images on the web. Indeed, this response is consistent with the Birmingham School tradition that McComiskey invokes. Raymond Williams, for instance, insists that critique is an incomplete response; "critical demystification can take us only part of the way" and should proceed "always in association with practice: regular practice, as part of a normal education [. . .]: practice in the production of alternative images of the 'same event'; practice in processes of basic editing and the making of sequences" (62).

Williams suggests a model of the public sphere in which rhetors appropriate the modes and media of the dominant culture to counter the practices of the consciousness industry. This model is consistent with recent discussions of the public sphere that emphasize the need for citizens to engage not just in dialogue, but in rhetorical practices whose aim is transforming the underlying cultural codes themselves. Scott Welsh, for instance, draws attention to "background culture," claiming that "cultural vocabularies" that naturalize key cultural concepts such as race and gender need to be changed (685). Drawing on the sophistic tradition, Welsh provides an alternative to deliberative exchanges that is characterized by

> political actors creatively interpreting and modifying commonly referenced or understood ways of speaking. The aim is seen not as attempting to change interlocutors' minds, one by one [. . .] but to effect a shift in prevailing relationships between and meanings of

> key cultural-political terms, events, or narratives. [...] [B]ackground culture becomes the "source" *and* "goal" of effective political speech governing meanings of a political collectivity. (690)

But, as González demonstrates, the cultural vocabularies that contextualize public discourse are multimodal vocabularies, comprised not just of words but of images as well as image-sound-word compounds. Key markers of race, class, gender, and sexual orientation are visual, as are markers of cultural locations, such as "inner city" and "suburb." Film and television contribute to the formation of cultural vocabularies by placing into circulation specific images that naturalize cultural constructions.

Welsh's emphasis on altering the signifying practices of the dominant culture parallels what Eco has usefully called "semiotic guerilla warfare" (qtd. in Hebdige 105). Eco refers to the ability of individuals and groups to appropriate existing semiotic materials and bend them to their own needs and purposes. Dick Hebdige deploys Eco's concept in *Subculture: the Meaning of Style*, observing that different subcultural groups use style rhetorically, bending the straight meaning of everyday objects to serve as tools of resistance:

> The limits of acceptable linguistic expression are prescribed by a number of apparently universal taboos. These taboos guarantee the continuing "transparency" (the taken-for-grantedness) of meaning. Predictably then, violations of the authorized codes through which the social world is organized and experienced have considerable power to provoke and disturb. (91)

Whereas traditional academic critique is about making visible "authorized codes" and the cultural work that they perform, semiotic guerilla warfare is about directly changing the semiotic fabric itself and thereby changing the consciousness that is shaped by this fabric.

Semiotic guerilla warfare accurately characterizes the multimodal practices in which many activist and community groups engage, and there is an emerging awareness within the field of composition, especially among feminists, that preparing students to represent their lived experiences multimodally should be a fundamental goal of rhetorical education. Gail Hawisher and Patricia Sullivan observe that "as women have more control over writing their own visualizations online, we see some women representing themselves complexly in creative, rhetorically effective ways" (288). They warn that "as inhabitants of this [visually saturated] world—as women, as English professionals, and as teachers—we cannot afford to ignore the visual. We do so at our own peril" (289). DeVoss and Selfe conclude,

> We are witnessing the emergence of new spaces for identity formation and display, spaces where women are rewriting conventional narratives of the public-private divide, the unified subject, and cyberspace as a male domain. New media and new realms have invited new rhetorical positionings for the creative souls working in these spaces, and as teachers of composition, we need to help students explore, develop, and communicate more effectively in them. (46)

The *Semiotics and the Media* website, produced collaboratively by Thomas Streeter and his students, offers an interesting example of semiotic warfare aimed at destabilizing gender codes. In their web essay "This is Not Sex," Streeter and his students restage a series of magazine advertisements, substituting men where women are depicted in the originals. The men mimic the poses and facial expressions of the original women models. The result is visually jarring, precisely because visual codes associated with specific cultural constructions of women—codes we normally take for granted—are graphically brought to our attention in the re-stagings. As the web essay proclaims, "Most viewers find the images of the men odd or laughable. But the images of the women seem charming and attractive." The visual re-stagings found in "This Is Not Sex" differ from traditional critique in that they act directly on the cultural codes themselves, disrupting them by applying to men visual conventions associated (in the dominant culture) with women.

Cultural locations like "inner city" can be reframed through appropriations of multimodal rhetoric as well. In *Teaching Youth Media*, for instance, Stephen Goodman writes about his work with urban teenagers at the Educational Video Center (EVC) in New York City. Goodman adopts an approach to media that "links media analysis to production; learning about the world is directly linked to the possibility of changing it" (3). Mindful that the "print and visual language that is used to name these kids and the worlds they inhabit can build public consent" for media practices that "frame" urban teens as "criminals and consumers," Goodman and the EVC create a learning environment in which teens produce and find public venues for showing documentary films that recode, reframe, and rename their identities and lived experiences in their own terms: "In the end, the young documentary makers constructed a powerful collection of words, music, and images that represented their own framing of reality" (29, 45). Likewise, David Sheridan explores the way grass-roots organizations in Detroit are using the Internet to place into circulation representations that more accurately reflect their lived experiences than those found in mainstream media.

These practices are suggestive of a public sphere that is consistent with Scott Welsh's notion of a space for revising and resisting practices of multimodal rhetoric associated with the dominant culture. The kinds of operations that can be performed on cultural codes have been described variously as "defamiliarizing," "disrupting," "interrupting," "subverting," "inverting," "bending," "signifying," and "jamming." While we do not wish to homogenize concepts that have been introduced by different individuals writing from within diverse contexts, all of these operations share the goal of undermining the "normal" operation of semiosis in the service of dominant ideology.

Rhetorical education has an important role to play in fostering this kind of public sphere participation. In order to take advantage of the special cultural work that multimodal rhetoric can perform, students, as public rhetors, need to be given opportunities to produce and repurpose multimodal compositions that counter existing hegemonic vocabularies. Doing this means confronting such questions as, What is the nature of the exigency that calls for an intervention? What rhetorical practices contribute to this problem? How is it reinforced and reproduced by the circulation of images, metaphors, stories, and representations? What kinds of counter-practices can effectively intervene? What new images, metaphors, stories, and representations need to be placed into circulation if consciousness is to be altered? How can hegemonic naming and framing practices be destabilized? How can new naming and framing practices be introduced?

NONSPECIALIST MEDIA PRODUCERS AND THE CHALLENGE OF INTERDISCIPLINARITY

Ultimately, when we consider the conditions that gave rise to Jim Ridolfo's creation of the FTAA film, we find ourselves returning to fundamental questions that present significant challenges for the contemporary academy: What curricular, structural, and infrastructural reconfigurations would enable people like Jim to "dive in" with new media technologies as they simultaneously develop an understanding of critical rhetoric? How should rhetorical education position itself in relation to other disciplines and fields that confront multimodality?

We have been encouraged by the increasing numbers of individuals who, like Jim, combine academically sanctioned learning with self-sponsored learning in order to produce effective multimodal compositions. These rhetors do not necessarily undergo the traditional credentialing processes of professional photographers, video editors, or graphic designers, but instead adopt a "learn-as-you-go" approach to multimodality. Embracing this practice, however, forces us to rethink the focus on disciplinary specialization that characterizes the contemporary academy. The current schema associates different forms of expression with different institutional sites, so that writing, graphic design, and music, are addressed in English, Art, and Music departments respectively, and redundancy is considered a form of inefficiency. In place of disciplinary structures that limit the use of new media technologies to specialists—structures that privilege "mastery" of technologies over social action—we envision a critical-praxis-based approach in which public rhetors appropriate all available means of rhetorical production. At times this may necessitate a kind of disciplinary trespass in which public rhetors and rhetorical educators cross into turf that has long been claimed by others in fields like studio art, television and video production, film studies, and music studies. We imagine a shrewdly pragmatic approach in which due respect is paid to disciplinary knowledge as long as it enables—not forestalls—activist appropriations of multimodality.

One can imagine, for instance, making introductory graphic design courses available to all undergraduates, giving them the opportunity to explore rich design principles like balance, rhythm, emphasis, and contrast. At the same time, critical praxis might demand a more expedient approach. An activist rhetor who has not taken a course in graphic design might make use, in response to a given exigency, of whatever is at hand, including approaches that professional designers would find reductive. She might turn to one of the many lists of "design tips" available on the web that offer formulas like "never use more than six words per line or six lines per slide." This strategic compromise is not ideal, but public rhetors do not operate in the realm of the ideal.

Those responsible for higher education, however, can design educational experiences so that the need for such compromises is limited by creating richer opportunities for interdisciplinary praxis. This means, in part, confronting the reality that higher education, as it is currently configured, suffers from a lack of sustained, formal opportunities for students to synthesize their discipline-specific learning experiences. We have distribution requirements, but few places where students are able to make connections between the discrete experiences that result from those requirements. Rhetorical education can function as a key site for interdisciplinary synthesis and critique, providing student citizens opportunities to interrogate the various disciplinary frameworks made available to them through discipline-specific experiences, to ask what practices those frameworks make possible and what practices they foreclose.

To accomplish this, rhetorical education needs to exploit what Rolf Norgaard calls "disciplinary contact zones"—institutional and intellectual sites "that place students at the margins of their own fields or that have them straddle organizational boundaries" (48). Synthesizing Pratt's notion of "social spaces where cultures meet, clash, and grapple with each other" with Bazerman and

Russell's "interface discourse" and Journet's "boundary rhetoric," Norgaard theorizes "educational experiences that foreground the negotiation of expertise" (49, 51). Disciplinary contact zones include learning structures located beyond and between individual courses: alternative learning spaces, multi-course portfolios, multidisciplinary teams, project- or problem-based learning, service learning, and residential and nonresidential learning communities.

A first-year engineering student, for instance, might enroll in a cluster of courses in writing, in graphic design, and in her major. In an approach that combines project-based learning with multi-course portfolios, she might be asked to identify a public exigency engineering might meaningfully address and then create a set of materials that will help a lay audience understand this problem and that will advocate for specific solutions. Over the course of a semester, she might produce a flyer, a tri-fold brochure, and a website—a coordinated series of deliverables that are linked by a coherent visual approach, that provide varying levels of detail, and that reflect kairotic assessments of modes and media. The student might be asked to document her learning in a holistic portfolio that includes these print materials in addition to a variety of developmental and reflective compositions (for example, analyses of target audiences and publication venues, plans for reproduction and distribution, analyses of what particular rhetorical strategies were used, estimates of effectiveness). In this scenario, the practices of rhetoric, design, and engineering are synthesized in a single project that foregrounds public goals. Meaningful interdisciplinarity is achieved by linking a set of discipline-specific experiences in ways that encourage critical reflection and integrated practice.

A different kind of contact zone is suggested by what Trimbur calls a "multiliteracy center": sites of collaborative learning in which student consultants support peers as they work on multimodal compositions ("Multiliteracies").[6] Multiliteracy centers go beyond computer labs staffed by technicians, offering support through consultants who have a sophisticated understanding of how rhetoric and technology are related and who have developed specialized skill sets through formal coursework, professional experience, and self-directed learning. The community institution that corresponds to the multiliteracy center is the community media center (CMC). As the Alliance for Community Media website explains, "For democracy to flourish, people must be active participants in their government [. . .]. Communications networks which use the public rights-of-way and public spectrum must provide the means and support for that participation through community uses of media." CMCs across the country are providing access to both the technologies and education necessary to engage in media production.

A critical praxis of multimodality ultimately requires a rethinking of rhetorical education that runs counter to the bottom-up tendency in the contemporary academy. This tendency assumes that it is only after students gain some exposure to ideas through "breadth" requirements will they be given access to the specialized knowledges necessary for them to do their job as professionals in the field. While this approach may make sense in terms of a vocational education designed to produce students with the technical skills (*techne*) necessary to run a camera, design photo layouts, or even produce a film, we are concerned that this approach reinforces the gap between technical expertise in new media technologies and social action.

BEYOND SNAP, CRACKLE, AND POP RHETORIC

Observing that "mass media leads a public to expect and to act on snappy one-liners," Arabella Lyon outlines several possible responses:

> We can theorize the effects of snap, crackle, and pop culture; we can produce more snap crackle and pop; we can theorize alternative public spaces; or we can develop and promote alternative public spheres and voices. [T]raditional theories [. . .] can only provide a starting point to any of these actions because the culture of late capitalism includes more voices, technologies, suspicions, and opportunities than the prior cultures where rhetoric has been theorized. (13)

In referring to mass media rhetoric as "snap crackle and pop," Lyon echoes a host of popular and academic critiques of mass media rhetoric that focus on its sound-bite nature and its preference for glitz and melodrama over substance. We have tried to acknowledge the power of mass media to shape the culture and the need to critique that power. But we have tried to go further, to imagine a culture that values a particular kind of rhetorical education aimed at preparing public rhetors to appropriate multimodal rhetoric for their own ends. We have examined the nature of this appropriation critically, however, viewing it as a complex endeavor characterized by both challenges and opportunities. Key to this project is the ability of multimodal rhetors to be critically reflective about the multimodal rhetorical strategies that will function most effectively within a given context. This means having access to the full range of the rhetorical tradition, including both confrontational and cooperative "moments." By mapping this tradition onto multimodality, we hope to sketch out a pedagogy that prepares students not to imitate, unreflectively, the rhetorical practices of mass media, but to go beyond snap, crackle, and pop rhetoric, transforming mass media practices strategically in ways that help facilitate social change.

NOTES

1. Following The New London Group, we use the term "multimodal" to refer to different "modes of meaning" that are combined in a single composition (25): a speech that combines gesture and spoken words; a digital slide show that combines graphs, written words, and spoken narration; a television show that combines moving images and a performed script. Although the distinction is not essential to our argument, we differentiate in this article between "mode" and "medium." The distinction is a slippery one, and a consensus about the precise meaning of these words has not yet developed. We use "media" to refer to mechanisms for delivering content to audiences. "Modes" refer to categories of semiotic content (which are based, in some scholarship, on the five senses). (See Kress and Van Leeuwen 21–22, Maybury and Wahlster 4–6, and Wyard and Churcher for discussions of mode and media).

2. A number of scholars have critiqued the narrow focus on the written word. George, Shipka, and Wysocki expose the artificial limits imposed by rhetorical approaches that dictate to students the mode, medium, and genre they must adopt. A different critique is implied by the work of Dyson, and Wade and Moje, who discuss the importance of the multiple literacies—including visual and multimodal literacies—that students bring with them when they enter the academy.

3. Apple Computer, for instance, has recently developed a suite of media production applications that are both cheap and easy to use. These applications are marketed, however, via images of leisure-time consumption that strongly echo the Kodak marketing Slater discusses: "What if you could command an entire world of music, photos, movies and DVDs—all from your sofa? Now you can share the good life with friends and family on a [. . .] new iMac G5" ("Mac Expo").

4. Shipka, drawing on Trimbur, rightly points out the importance of integrating delivery into our pedagogies, but again her framework needs to be transformed and extended if it is to effectively serve public rhetoric. One student, for instance, produces an intricate composition comprised in part by a set of specialized mirrors. Although this is clearly a provocative approach to the problem of delivery, it does not seem to be informed by a broader understanding of circulation as a process that extends beyond the classroom, eliding such questions as, How could this intricate composition be effectively reproduced so that a sufficient number of copies could be made to address the exigency? How could they be distributed to their intended audience? What are the costs involved in reproduction and distribution? How do the benefits of this ap-

proach weigh against competing approaches that might be easier to distribute?

5. Anne Wysocki has suggested redefining "new media" in terms that can be described as "self reflexive" in the sense we adopt for the term here: "I think we should call 'new media texts' those that have been made by composers who are aware of the range of materialities of texts and who then highlight the materiality: such composers design texts that help readers/consumers/viewers stay alert to how any text—like its composers and readers—doesn't function independently of how it is made and in what contexts. Such composers design texts that make as overtly visible as possible the values they embody" (15).

6. Technology-rich spaces that approach the kind of multiliteracy center Trimbur imagines are exemplified in Michigan State University's Writing Center, Michigan Tech's Center for Computer-Assisted Language Instruction, and Clemson University's Class of 1941 Studio.

WORKS CITED

Allen, Nancy. "Ethics and Visual Rhetorics: Seeing Is Not Believing Anymore." *Technical Communication Quarterly* 5.1 (1996): 87–105

Alliance for Community Media. "Dossier: Cable—Alliance for Community Media Public Policy Platform." *Community Media Review*, Spring 2005. 7 Dec. 2005. http://www.cctv.org/cmracm.htm.

Asen, Robert. "Toward an Normative Conception of Difference in Public Deliberation." *Argumentation and Advocacy* 35 (1999): 115–29.

Barthes, Roland. "Rhetoric of the Image." Trans. Stephen Heath. *Image, Music, Text*. New York: Hill, 1988.

Bazerman, Charles, and David R. Russell. *Landmark Essays on Writing Across the Curriculum*. Davis, CA: Hermagoras, 1994.

Bitzer, Lloyd. "The Rhetorical Situation." *Philosophy and Rhetoric* 1.1 (1968): 1–14.

Blair, Carole. "Contemporary U.S. Memorial Sites as Exemplars of Rhetoric's Materiality." *Rhetorical Bodies*. Ed. Jack Selzer and Sharon Crowley. Madison: U of Wisconsin P, 1999. 16–57.

Booth, Wayne C., and Marshall W. Gregory. *The Harper and Row Rhetoric: Writing as Thinking, Thinking as Writing*. New York: Harper, 1988.

Burgin, Victor. "Looking at Photographs." *Thinking Photography.* Ed. Victor Burgin. London: Macmillan, 1982. 142-53.

Calhoun, Craig J. *Habermas and the Public Sphere. Studies in Contemporary German Social Thought.* Cambridge: MIT P, 1992.

Corbett, Edward P.J. "The Rhetoric of the Open Hand and the Rhetoric of the Closed Fist." *College Composition and Communication* 20 (1969): 288-96.

Cornyn, Alison, et al. *360degrees: Perspectives on the U.S. Criminal Justice System.* 1 February 2006 http://360degrees.org/.

DeLuca, Kevin Michael. *Image Politics: The New Rhetoric of Environmental Activism.* New York: Guilford, 1999.

DeVoss, Dànielle N. "Women's Porn Sites—Spaces of Fissure and Eruption or 'I'm a Little Bit of Everything.'" *Sexuality and Culture* 6.3 (2002): 75-94.

DeVoss, Dànielle, Ellen Cushman, and Jeff Grabill. "Infrastructure and Composing: The *When* of New-Media Wrtiting." *College Composition and Communication* 57 (2005): 14-44.

DeVoss, Dànielle N., and Cynthia L. Selfe. "'This Page Is under Construction': Reading Women Shaping On-Line Identities." *Pedagogy: Critical Approaches to Teaching Literature, Language, Composition, and Culture* 2.1 (2002): 31-49.

Dragga, Sam. "The Ethics of Delivery." *Rhetorical Memory and Delivery: Classical Concepts for Contemporary Composition and Communication.* Ed. John Frederick Reynolds. Hillsdale: Erlbaum, 1993. 79-95.

Dyson, Anne Haas. *The Brothers and Sisters Learn to Write: Popular Literacies in Childhood and School Cultures.* New York: Teachers College P, 2003.

Eco, Umberto. "Critique of the Image." *Thinking Photography.* Ed. Victor Burgin. London: Macmillan, 1982. 32-38.

———. "Towards a Semiotic Enquiry into the Television Message." *Working Papers in Cultural Studies* 3, U of Birmingham, 1972.

Ellertson, Anthony. "Some Notes on Simulacra Machines, Flash in First-Year Composition, and Tactics in Spaces of Interruption." *Kairos* 8.2 (2003). 12 January 2005 http://english.ttu.edu/kairos/8.2/binder.html?features/ellertson/home.html.

Ellertson, Anthony, and Margaret Baker Graham. "In the Cyberhood: Using Global Technologies for Local Purposes." *Computers and Composition Online* (Fall 2003). 12 Jan. 2005 http://www.bgsu.edu/cconline/ellerston-graham/cyberhood.htm.

Enzensberger, Hans Magnus, and Michael Roloff. *The Consciousness Industry; on Literature, Politics and the Media.* New York: Seabury, 1974.

Faigley, Lester. "Material Literacy and Visual Design." *Rhetorical Bodies.* Ed. Jack Selzer and Sharon Crowley. Madison: U of Wisconsin P, 1999. 171–201.

Finnegan, Cara A. "Doing Rhetorical History of the Visual: The Photograph and the Archive." *Defining Visual Rhetorics.* Ed. Charles A. Hill and Marguerite Helmers. Mahwah, NJ: Erlbaum, 2004. 87–110.

Fulkerson, Richard. *Teaching the Argument in Writing.* Urbana: NCTE, 1996.

Garnham, Nicholas. "The Media and the Public Sphere." *Habermas and the Public Sphere.* Ed. Craig J. Calhoun. Cambridge: MIT P, 1992. 421–80.

George, Diana. "From Analysis to Design: Visual Communication in the Teaching of Writing." *College Composition and Communication* 54 (2002): 11–38.

Goggin, Maureen Daly. "Visual Rhetoric in Pens of Steel and Inks of Silk: Challenging the Great Visual/Verbal Divide." *Defining Visual Rhetorics.* Ed. Charles A. Hill and Marguerite Helmers. Mahwah, NJ: Erlbaum, 2004: 87–110.

González, Jennifer. "Morphologies: Race as a Visual Technology." *Only Skin Deep: Changing Visions of the American Self.* Ed. Coco Fusco and Brian Wallis. New York: International Center of Photography in association with H.N. Abrams, 2003. 379–93.

Goodman, Steven. *Teaching Youth Media: A Critical Guide to Literacy, Video Production, and Social Change.* New York: Teachers College P, 2003.

Graff, Gerald. "Two Cheers for the Argument Culture." *Clueless in Academe: How Schooling Obscures the Life of the Mind.* New Haven: Yale UP, 2003.

Haas, Christina, and Christine M. Neuwirth. "Writing the Technology That Writes Us: Research on Literacy and the Shape of Technology." *Literacy and Computers: The Complications of Teaching and Learning with Technology.* Ed. Cynthia Selfe and Susan Hilligoss. New York: MLA, 1994. 319–35.

Habermas, Jürgen. *The Structural Transformation of the Public Sphere: An Inquiry into a Category of Bourgeois Society.* Trans. Thomas Burger and Frederick Lawrence. Cambridge: MIT P, 1989.

Hands, Joss. "E-Deliberation and Local Governance: The Role of Computer Mediated Communication in Local Democratic Participation in the Unitized Kingdom." *First Monday.* 10.7 (2005): 28 Jan. 2006. http://www.firstmonday.org/issues/issue10_7/.

Hawisher, Gail E., and Patricia A. Sullivan. "Fleeting Images: Women Visually Writing the Web." *Passions, Pedagogies, and 21st Century Technologies.* Ed. Gail E. Hawisher and Cynthia L. Selfe. Urbana: NCTE, 1997. 268-91.

Hebdige, Dick. *Subculture: the Meaning of Style.* London: Methuen, 1979.

Hess, Douglas D. "2005 CCCC Chair's Address: Who Owns Writing?" *College Composition and Communication* 57 (2005): 335-57.

Hill, Charles A. "The Psychology of Rhetorical Images." *Defining Visual Rhetorics.* Ed. Charles A. Hill and Marguerite Helmers. Mahwah, NJ: Erlbaum, 2004: 25-40.

Hine, Lewis. "Social Photography." *Classic Essays on Photography.* Ed. Alan Trachtenberg. New Haven: Leetes, 1981. 109-13.

Ivie, Robert. "Rhetorical Deliberation and Democratic Politics in the Here and Now." *Rhetoric and Public Affairs.* 5.2 (2002): 277-85

Jarratt, Susan C. "Feminism and Composition: The Case for Conflict." *Contending with Words: Composition and Rhetoric in a Postmodern Age.* Ed. Patricia Harkin and John Schilb. New York: MLA P, 1991. 105-23.

Journet, Debra. "Interdisciplinary Discourse and 'Boundary Rhetoric': The Case of S.E. Jelliffe. *Written Communication* 10 (1993): 510-41.

Jenkins, Henry, and David Thorburn, eds. *Democracy and New Media.* Cambridge: MIT P, 2003.

Kienzler, Donna S. "Visual Ethics." *Journal of Business Communication* 34.2 (1997): 171-87.

Kinneavy, James L. "*Kairos*: A Neglected Concept in Classical Rhetoric." *Rhetoric and Praxis: The Contribution of Classical Rhetoric to Practical Reasoning.* Ed. Jean Dietz Moss. Washington, D.C.: Catholic U of America P, 1986. 79-105.

Kress, Gunther R., and Theo Van Leeuwen. *Multimodal Discourse: The Modes and Media of Contemporary Communication.* London: Arnold, 2001.

Laclau, Ernesto, and Mouffe, Chantal. *Hegemony and Socialist Strategy.* Trans. Winston Moore and Paul Cammack. London: Verso, 1985.

Lanham, Richard A. *The Electronic Word: Democracy, Technology, and the Arts.* U Chicago P, 1993.

Lynch, Dennis A., Diana George, and Marilyn M. Cooper. "Moments of Argument: Agonistic Inquiry and Confrontational Cooperation." *College Composition and Communication* 48 (1997): 61–85.

Lyon, Arabella. *Intentions: Negotiated, Contested, and Ignored.* University Park: Pennsylvania State UP, 1998.

"Mac Expo 2005." 3 Feb. 2006 http://www.mac-expo.co.uk/.

Macrorie, Ken. "Blow That Horn, Man." *College Composition and Communication* 14 (1963): 215–19.

Maybury, Mark T., and Wolfgang Wahlster. "Intelligent User Interfaces: An Introduction." *Readings in Intelligent User Interfaces.* Ed. Mark T. Maybury and Wolfgang Wahlster. San Francisco: Kaufmann, 1998.

McChesney, Robert Waterman. *Rich Media, Poor Democracy: Communication Politics in Dubious Times.* Urbana: U of Illinois P, 1999.

McComiskey, Bruce. Rev. of *Rhetorical Bodies* Ed. by Jack Selzer and Sharon Crowley. *JAC* 20 (2000): 699–703.

——. "Social-Process Rhetorical Inquiry: Cultural Studies Methodologies for Critical Writing About Advertisements." *JAC* 17 (1997): 381–400.

——. "Visual Rhetoric and the New Public Discourse." *JAC* 24 (2004): 187–206.

Messaris, Paul. *Visual Persuasion: The Role of Images in Advertising.* Thousand Oaks, CA: Sage, 1997.

Monroe, Barbara. *Crossing the Digital Divide: Race, Writing, and Technology in the Classroom.* New York: Teachers College P, 2004.

Mouffe, Chantal. "For an Agonistic Model of Democracy." *The Democratic Paradox.* London: Verso, 2000: 80–107.

The New London Group. "A Pedagogy of Multiliteracies: Designing Social Futures." *Multiliteracies: Literacy Learning and the Design of Social Futures.* Ed. Bill Cope and Mary Kalantzis. London: Routledge, 2000. 9–37.

Norgaard, Rolf. "Negotiating Expertise in Disciplinary 'Contact Zones.'" *Language and Learning Across the Disciplines* 3.2 (1999): 44–63.

Pendakur, Manjunath, and Roma Harris, eds. *Citizenship and Participation in the Information Age.* Aurora, Ontario: Garamond P, 2002.

Pratt, Mary Louise. "Arts of the Contact Zone." *Profession 91.* New York: MLA: 33–40.

Ruby, Jay. *Picturing Culture: Explorations of Film and Anthropology.* Chicago: U of Chicago P, 2000.

Selber, Stuart A. "Reimagining the Functional Side of Computer Literacy" *College Composition and Communication* 55 (2004): 470–503.

Jack Selzer, and Sharon Crowley, eds. *Rhetorical Bodies.* Madison: U of Wisconsin P, 1999. 16–57.

Shapiro, Michael J. *The Politics of Representation: Writing Practices in Biography, Photography, and Policy Analysis.* Madison: U of Wisconsin P, 1988.

Sheridan, David M. "Digital Detroit and the Frail Particulars of Everyday Life." *Journal of Literacy and Technology* 2.1 (2002). 12 Jan. 2005 http://www.literacyandtechnology.org/v2n1/sheridan.html.

Shipka, Jody. "A Multimodal Task-Based Framework for Composing." *College Composition and Communication* 57 (2005): 277–306.

Slater, Don. "Marketing Mass Photography." *Visual Culture: The Reader.* Ed. Jessica Evans and Stuart Hall. London: Sage, 1999. 289–306.

Sontag, Susan. *On Photography.* New York: Doubleday, 1990.

Streeter, Thomas, et al. "This Is Not Sex: A Web Essay on the Male Gaze, Fashion Advertising, and the Pose." 12 Jan. 2005 http://www.uvm.edu/~tstreete/powerpose/.

Stroupe, Craig. "Visualizing English: Recognizing the Hybrid Literacy of Visual and Verbal Authorship on the Web." *College English* 62 (2000): 607–32.

Tagg, John. *The Burden of Representation: Essays on Photographies and Histories.* Minneapolis: U of Minnesota P, 1993.

Tannen, Deborah. *The Argument Culture: Moving from Debate to Dialogue.* New York: Random, 1998.

Tan, Ed S. *Emotion and the Structure of Narrative Film: Film as an Emotion Machine.* Hillsdale, NJ: Erlbaum, 1996.

Trimbur, John. "Composition and the Circulation of Writing." *College Composition and Communication* 52 (2000): 188-219.

———. "Delivering the Message: Typography and the Materiality of Writing." *Visual Rhetoric in a Digital World: A Critical Sourcebook*. Ed. Carolyn Handa. Boston: Bedford, 2004.

———. "Multiliteracies, Social Futures, and Writing Centers." *Writing Center Journal* 20.2 (2000): 29-32.

Wade, Suzanne E., and Elizabeth B. Moje. "The Role of Text in Classroom Learning." *Handbook of Reading Research, Vol. III*. Ed. Michael L. Kamil et al. Mahwah, NJ: Erlbaum, 2000. 609-27.

Ward, Irene. "How Democratic Can We Get?: The Internet, the Public Sphere, and Public Discourse." *JAC* 17 (1997): 365-79.

Weisser, Christian R. *Moving Beyond Academic Discourse: Composition Studies and the Public Sphere*. Carbondale: Southern Illinois UP, 2002.

Welch, Kathleen E. *Electric Rhetoric: Classical Rhetoric, Oralism, and a New Literacy*. Cambridge: MIT P, 1999.

Welsh, Scott. "Deliberative Democracy and the Rhetorical Production of Political Culture." *Rhetoric and Public Affairs* 5.4 (2002): 679-708.

Williams, Raymond. *Problems in Materialism and Culture: Selected Essays*. London: Verso, 1980.

Wyard, Peter, and Gavin Churcher. "All Channels Open: Multimodal Computer Interfaces." *BT Technical Journal* 18.1 (2000). 12 Jan. 2005 http://more.btexact.com/millennium_issue/vol18no1/today.htm.

Wysocki, Anne Francis. "Opening New Media to Writing: Openings and Justifications." *Writing New Media: Theory and Applications for Expanding the Teaching of Composition*. Ed. Anne Francis Wysocki, et al. Logan: Utah State UP, 2004. 1-41.

Young, Iris Marion. "Communication and the Other: Beyond Deliberative Democracy." *Democracy and Difference: Contesting the Boundaries of the Political*. Ed. Seyla Benhabib. Princeton: Princeton UP, 1996.

5

THE POLITICAL ECONOMY OF COMPUTERS AND COMPOSITION

"DEMOCRACY HOPE"
IN THE ERA OF GLOBALIZATION

M.J. Braun

In a review of recent critical work in the field of computers and composition, Ken McAllister writes, "By and large, the history of the incorporation of computers into educational settings has been oppositional, the technophiles and the technophobes exchanging verbal barbs with too little attention paid to history and economics" (192). In academic journals, the opposition has been too often depicted as conflicting identities: on one side, the luddites (driven by modernist, humanist, or Marxist idealism) resist the brave new computerized world; on the other side, the cyborgs (inspired by theories of *différance*, language games, and simulacra) dive head first into the electronic stuff of our postmodern era. Both depictions are harmfully reductive because they ignore the historical complexity of each identity. The luddites of the early nineteenth century were much more than machine-smashers; their war on technology ensued only after their attempts to intervene democratically in economic and political agendas were frustrated (see Dinwiddy; Peel; Sale). The cyborg, as constructed by science-fiction writers and academic critics of scientific rationalism, experiences its machine/flesh condition through its implication in a multiplicity of political and economic contradictions (see Haraway). Luddites were not politically naive after all, and neither are cyborgs uncritically assimilated into the new economy.

Ahistorical and reductive depictions of luddites and cyborgs in scholarly literature help to reproduce the verbal barbs that McAllister cautions us against. However, in his review, he cites recent thinking that attempts to resist the reductive luddite-cyborg binary by taking into consideration the economic and political contexts of technologized pedagogy and the new technologized identities of the students who enter our classrooms. In this article, I add to this critical trend.

Specifically, I analyze the pedagogies and proposals for political action laid out by Chris Anson, James Berlin, David Downing, Charles Moran, and Cynthia Selfe, and I argue that each of their positions transcends the reductive binary and that each is aware that our classrooms and institutions are historically situated in economic and political structures. Of course, by stating that I will add to this work, I am also implying that something is lacking there. What is missing is the critique of capital, the systemic backdrop against which all discussions of identity, history, and politics must be staged. Consequently, I examine the technologizing of composition through the critical lens of political economy, a field of inquiry begun in the late seventeenth and early eighteenth centuries that investigates how wealth accumulates in capitalist economies.

Since the mid-nineteenth century, most analysis of political economy has derived from a Marxist orientation—that is, from the viewpoint of those who do not benefit from capitalist accumulation. Following Marx, I understand capital to be the dialectical identity of macro-systems and micro-practices. At the macro-level, capital embodies the absolute drive to create surplus value (profit) and has historically produced surplus value through four macro-systems: mercantilism, competitive capitalism, monopoly and corporate capitalism (imperialism), and global capitalism. Although each macro-structure predominates at different times in particular places, all of these forms can and do coexist, as Marx repeatedly argues: "Epochs in the history of society are no more separated from each other by strict and abstract lines of demarcation than are geological epochs" (492). At the micro-level, in contrast, capital embodies myriad, relative practices through which profit is generated: the creation of manufacturing systems to produce commodities for exchange; the creation of proletarianized people whose labor can be exploited; and the creation of state forms that enable these practices. Each of these components assumes a particular material form that is flexible and fluid (the phenomenon that Marx and Engels referred to in *The Communist Manifesto* as capital's dynamic character); however, the drive to exploit for profit is absolute, inflexible, and foundational to capital. It is crucial, therefore, that we distinguish between absolute and relative surplus value when we discuss new technologies because technology only appears as relative surplus value; its appearance is temporary, ephemeral, and incidental. However, any particular technology, as with all forms of relative surplus value, is always identified in its dialectical relationship with its opposite: the absolute need to generate surplus value.

My contribution to the current body of critical work also leads me to different and somewhat contradictory conclusions than four of the five authors I review. Berlin and Downing, although working from very different theoretical frameworks, support the premise that new technologies enable methods of work that help create a new democratic subject: the citizen-worker.[1] By importing new technologies into composition pedagogy, teachers can nurture the new democratic consciousness of the citizen-worker among their students. Moran and Selfe, in contrast, support the premise that there are no inherently democratizing characteristics or effects of new technology. Rather, technologies emerge within the social divisions already at work in the U.S. economy and culture. Therefore, both urge teachers, students, and members of local communities to use their positions as democratic subjects to lobby for equal access to new technologies for those currently excluded from it. Anson, however, premises his position on the logic of the capitalist economy—its drive for efficiency. This recognition leads him to propose political action substantially different from Berlin, Downing, Moran, and Selfe. He argues that political resistance should target the capitalist logic of efficiency.

I argue that the positions taken in each essay, with the exception of Anson's position, share uninterrogated assumptions about capitalism. Specifically, they assume that it only exists as micro-practices not as a macro-system, that these capitalist micro-practices are not antithetical to the expansion of democracy, and that social antagonisms can be resolved through the regulation rather than the elimination of capital. These beliefs together constitute an ideology I call "democracy hope," an ideology that always leads to the false promise that, if we just work hard enough for and through democracy, social inequality can be mitigated. As with all ideologies, democracy hope allows us to see certain things while blinding us to others. It allows us to see capital's micro-practices, such as changes in forms of work or types of technologies, but prevents us from recognizing that capital is also a macro-system that mobilizes all of its micro-practices according to its drive for profit. Democracy hope allows us to recognize the existence of social divisions, but it prevents us from seeing how they are constructed as class antagonisms. In addition, it allows us to see the relative democratic privileges of the middle class, but it prevents us from seeing how these privileges are only realized through the exploitation, immiseration, and global expansion of proletarianized people.

To establish a basis for defining democracy hope and for offering a way out of it, I first explore Marx's theory of absolute and relative surplus value. Next, I turn to an academic field of inquiry—the political economy of communication—that uses a Marxist theoretical framework to analyze twentieth-century media and communication technologies. Because our own field lacks a thoroughgoing Marxist methodology, the political economy of communication offers us some instructive models for critique.[2] Finally, I argue that because the field of computers and composition has not yet rigorously considered the Marxist critique of capital, the ideology of democracy hope persists and distorts what we see when we look critically at the political implications of our practices with technology. This ideology reproduces the idea that some kind of social leveling can be achieved by our own democratic practice with technology through consciousness-raising and collaboration, expanded access to technologies, and democratic engagement with institutional, corporate, or state agendas. I concur that we should act at these sites, but to contribute to the struggle for social and economic justice for those at the bottom of the social hierarchy, we must refashion our discourse. We must expose the class antagonism at work. I conclude that we must name names and signify capital as the exploitative system that it is; in order to do this, we must shed assumptions about the role of antagonism in argumentation. We must be aware that as long as class formations are regimented by the macro-system of capital— its absolute drive to produce surplus value—democracy may be expanded to the middle class, but it will remain unavailable to proletarianized people.

CLASSICAL POLITICAL ECONOMY AND MARX'S CRITIQUE

When questioned why he never cited Marx in his work, Michel Foucault responded that attributing historiographic concepts and methods to Marx is as unnecessary as citing Einstein in discussions of relativity:

> It is impossible at the present time to write history without using a whole range of concepts directly or indirectly linked to Marx's thought and situating oneself within a horizon of thought which has been defined and described by Marx. One might even wonder what difference there could ultimately be between being a historian and being a Marxist. ("Prison" 53)

Foucault's point seems even more apropos if the historical inquiry undertaken is political economy, a field popularized in the West by eighteenth-century Scottish philosophers such as Adam Smith. From its inception, Vincent Mosco explains, this field merged three areas of knowledge—politics, economics, and ethics—and considered its work as praxis rather than pure theory. The discipline of economics, in contrast, was conceptualized and codified, along with many other disciplines at the end of the nineteenth century, to follow the empirical model of positivist science and has restricted itself to quantitative inquiry (63-68). The creation of economics as a university-based discipline pushed political economy, with its tripartite structure, into the streets, where it informed the praxis of radical trade unionist, anarchist, socialist, and communist parties. Robert McChesney argues that when political economy was brought back to the academy by communications scholars, the field retained its traditional tripartite roots, but it was overwhelmingly dominated by Marx's reconceptualizations of it (3–12). Economics and political economy, therefore, are separate disciplines with distinct objects of inquiry. Mosco's and McChesney's claims about the impossibility of writing political economy outside of Marx's thought echoes Foucault's claim about writing history. In this section, I present a brief overview of Marx's theoretical break with classical political economy and argue that the Marxist "horizon of thought," within which political economists have since situated themselves, emerges from Marx's reworking of the Hegelian dialectic and his application of it to the concept of surplus value that is first developed in classical political economy.

In *Reading Capital*, Louis Althusser and Étienne Balibar explain that classical political economy sought to determine how wealth accumulates under capitalism. Eighteenth-century political economists witnessed the moment of nascent capitalist production and exchange and were in awe of the speed and ubiquity of its effects on the total social order. They noted that wealth was beginning to be acquired in previously inconceivable ways. They correctly theorized that the accumulation of wealth characteristic of feudalism—a very visible, quantitative acquisition of land and resources—was beginning to be replaced by a more invisible and dynamic accumulation brought about by capitalist mercantile and manufacturing practices. In this new form, wealth seemed to be created rather than accumulated; they noticed that a value in excess of what could be physically quantified and visibly identified existed at the end of the production/exchange process that did not exist when the process began. This value came to be called surplus value.

Althusser and Balibar note that Marx, writing a century after Smith, witnessed a different moment in the history of capitalist production: the shift from craft/workshop methods of production to large-scale machine-based industry. The shift set the stage for Marx to break radically with classical political economy. In the industrial factory, hundreds of workers were socially organized; their work practices were dramatically changed from crafting commodities in workshops (which required both mental and manual labor) to manipulating machines, which removed the workers' minds from the process of crafting and disciplined their hands and bodies in increasingly repetitive and deskilled regimes of mind-numbing motions. Because classical political economists lived prior to the invention of mass production, they remained influenced by the quantitative accumulation model, even though they sought to discover the new forms in which capital creates wealth. They identified three sources of surplus value: value derived from labor, value derived from property (in the extraction of rent), and value derived from capital itself (in the extraction of interest). Althusser and Balibar explain that because Marx witnessed the production of surplus value in large-scale industry, he was able to problematize the accumulation of wealth. At the level of the problematic, old disciplinary frameworks break apart and new theory develops that redefines, as Foucault argues, what falls "within the true" ("Discourse" 224). In his break with classical political economy, Marx theorizes that all surplus value, regardless of the appearance of its source (be it profit, rent, or interest) is the product of human labor. For Marx, human activity alone creates value, and, more profoundly, human consciousness creates the concept of value itself.

Althusser and Balibar argue that by introducing the problematic, Marx changes the object of inquiry of political economy from the collection of empirical data regarding the accumulation of wealth to the functioning of surplus value itself (147–57). Marx applies his reconception of Hegel's dialectics to this new object of inquiry, allowing him to divide surplus value into two. He theorizes that surplus value is not an economic fact; rather, it is a relationship that only exists as the dialectical tension of its two forms: the universally abstract form (absolute surplus value) and the physically concrete and constantly changing form (relative surplus value).

Mosco contends that Marx's application of the dialectic to classical political economy represents a shift not only in his economic vision, but in his political and ethical vision as well. The decision of classical political economists, he argues, "to define labor along with land and capital" as the source of surplus value "also reflects a certain moral vision, however implicit, that people are interchangeable with capital" (36). Marx, in contrast, shows how human beings are the sole agents of making history "albeit under conditions that are not of their own making" (Mosco 44). By specifiying labor as the sole creator of value in all its concrete and abstract forms, Marx is able to identify the historical emergence of capital as a new type of social relation. Marx consciously uses the word *capital*, rather than *capitalism*, to refer to this new social relation because *capitalism* implies that the new method of producing surplus is merely a preferred way of generating surplus value, not a foundational practice of the new economic system. As a social relation, capital raises new political and ethical problems, the heart of which is its foundational reliance on the existence of a new class, the proletariat, from whom surplus value is directly extracted. In *The Communist Manifesto*, Marx and Engels define this class as "laborers, who live only so long as they find work and who find work only so long as their labor increases capital . . . who must sell themselves piecemeal, are a commodity, like every other article of commerce, and are consequently exposed to all the vicissitudes of competition, to all the fluctuations of the markets" (29). The proletariat is constructed in the same way the products of its labor are: as a commodity to be traded in the marketplace. Its value is set abstractly, not concretely. Proletarians are not paid by the concrete hours or even by the various concrete skills they bring to the workplace. Rather, as a commodity on the market, the price of proletarian labor is set according to the absolute minimum necessary for this class to sustain its ability to work and its ability to stay alive and produce the next generation of workers. The implications of the foundational role of the proletariat in Marx's theory of value cannot be overemphasized: as long as there is capital, there is also a class of people who must live at a minimal level of subsistence—a level Marx called "immiseration." This foundational principle of the Marxist theory of value is often completely overlooked or rejected outright by academic Marxists. In my view, this principle still holds, and it will hold as long as capital exists as the definitive social relation. Furthermore, this principle is central to my argument against democracy hope.

In sum, many different relative class formations and technologies may come and go, but the production of proletarianized people is required in order to produce surplus value in each of its increasingly consolidated forms: mercantilism, competitive capitalism, imperialism, and global capitalism. Each new technique, each new technology, each new class formation does not bring an end to the constant production of the proletariat. In *Capital*, Marx chillingly captures nineteenth-century factory life and the role of new technologies in creating this life: the mutilation of bodies tied to machines, the severing of the mind from labor, and the sacrifice of human life on the altar of the market. These conditions, I might add, are presently proliferating in the Third World, as well as inside the capitalist centers in the post-Cold War "globalized" economy.

I have described the dialectic between absolute and relative surplus value because Berlin's and Downing's positions promise that a potentially flourishing democracy may emerge from the micro-practices of new technologies and their associated regimes of work. Based on my reading of *Capital*, however, I argue that this is an impossible hope. As relative surplus value, technologies and the changing class formations they generate do not act independently of capital's absolute drive for surplus labor (Marx 432, 645). The democratizing effect of new technologies in one sector of the economy must be accompanied by the further imposition of tyrannical control over workers' lives in other sectors. Otherwise, surplus value cannot be extracted from human labor. For the purposes of analyzing Anson's, Moran's, and Selfe's positions, it is more helpful to turn to Marx's discussion of the relationship between the capitalist class and its state, and the attempts of the managerial and professional middle classes to allay the degradation of the proletarian class through democratic state forms. Moran and Selfe seek reform of administrative and governmental educational policies at the local, state, and national level. They want the growing gap between the technological haves and have-nots—the phenomenon the Department of Commerce has labeled the "digital divide"—to be closed.[3] Moran and Selfe propose that teachers, students, and others become involved in the democratic process and focus their arguments on the need to lessen the digital divide in education. Again, based on my reading of *Capital*, I argue that Moran and Selfe also harbor an impossible hope. Anson, however, does not. He also advocates that political action be taken, but he does not focus his attention on reforming the digital divide. Instead, he argues that political action should expose the logic of efficiency behind the rapid technologizing of education. The distinction between these two political positions—reform of capitalism versus exposure of capitalism—suggests ways to avoid democracy hope.

Marx offers a way of understanding this distinction in his discussion of the role of the state in maintaining the interests of the capitalist class in the face of middle-class movements to reform the egregious conditions of proletarianized life. One of the social reform movements in Marx's day emerged from a public outcry in England over child labor and the rising death rate among working-class women and their children. The movement that coalesced against this degradation in the early nineteenth century instigated the passage of the British Factory Act in 1844, which purported to regulate child labor. Marx analyzes the social effects of this reform on the proletariat by comparing the conditions of women's and children's lives before and after the Factory Act was passed:

> Before the labour of women and children under ten years old was forbidden in mines, the capitalists considered the employment of naked women and girls, often in company with men, so far sanctioned by their moral code, and especially by their ledgers, that it was only after the passing of the Act that they [British capitalists] had recourse to machinery. The Yankees invent a stone-breaking machine. The English do not make use of it because the "wretch" who does this work gets paid for such a small portion of his labour that machinery would increase the cost of production to the capitalist. In England women are still occasionally used instead of horses for hauling barges, because the labour required to produce horses and machines is an accurately known quantity, while that required to maintain the women of the surplus population is beneath all calculation. (516–17)[4]

Marx is making two important points here about reform movements that emanate from the professional and middle classes. First, the drive for absolute surplus value always delimits the technologies and class formations that appear as relative surplus value. In his example, mine owners only incorporate new stone-breaking machines into the production process once it becomes more profitable to exploit the machinery rather than to exploit the bodies of women and children. Specifically, mine owners deemed those machines to be profitable only when a law curtailed the older, more directly exploitative forms of extracting surplus value from workers' bodies. New technologies, therefore, may be invented, but they are only used once their exploitation becomes cheaper than the exploitation of human labor, or once the state intervenes in the regulation of industry.

For Marx, legislation, like new technology, is nothing more than a technique of relative surplus value; it is in thrall to the absolute drive for surplus value. The Factory Act was supposed to end the kind of degradation that was common in mining—at least for children. But as he traced the twenty-three year interim between the passage of the Factory Act to the publication of *Capital Volume One* in 1867, Marx found that the degradation of children did not change—only the particular form of degradation changed (517-26). Marx poured over scores of documents written by factory inspectors (the nineteenth-century equivalent of OSHA), studies by professionals in medicine and education, records of governmental commission hearings, and parliamentary debates over proposed legislation. He discovered that, indeed, more children, in accordance with the Act, were attending school; however, because the Act did not abolish child labor but only restricted it by requiring laboring children in selected industries to attend at least three hours of school per day, children's lives became more defined by the needs of capital. The school system established by the law provided education seasonally, in accordance with an industry's production schedule rather than with a daily consistency that allowed for retention of what was learned. In addition, most of the schools established were only capable of providing a place to store children, since they were no more than small rooms stuffed wall-to-wall with children's bodies and, often, with "instructors" who could neither read nor write. The Factory Act benefitted the image of the capitalists, who appeared to support the education of the children they employed, but it failed to provide any real educational benefit to those children.

Another effect of the Factory Act was that infant mortality skyrocketed. By 1864, twenty years after the Act was passed, the "Sixth Report on Public Health" referred to Dr. Henry Julian Hunter's 1861 study entitled, "Excessive Mortality of Infants in Some Rural Districts of England" (Marx 521-22). Dr. Hunter's report concluded that higher death rates were due to the fact that more mothers were working in factories to make up for the diminution of total family income created by the Act's restriction of the hours that children could work. The absence of mothers from the home left no one to care for their infants. Older siblings who may have cared for these infants were also less available because they were farmed out by their parents to work all day in the industries that did not fall under the Act's schooling requirement. Dr. Hunter noted that desperate women who could not care for their babies turned to infanticide by opiates as the only available means of dealing with their new predicament. Marx concludes that although many from the middle and professional classes demanded reforms of the factory system, the type of reforms produced by the state enabled the capitalist class to continue producing surplus value by merely shifting the burden of producing it from one section of the working class to another. This observation must be taken to heart by all well-intentioned professionals who presently struggle for state reform of capitalist micro-practices. The double jeopardy that proletarians face after reforms become realized in law is commonplace in the contemporary discourse of democratic reform movements. Contemporary reformers often observe that once having obtained the reform they demanded from the state, new, unforeseen, and unwanted consequences result. For the proletarian class on behalf of whom the struggle for reform was fought, the consequence of reform is as detrimental on the whole as the situation prior to reform; for the reformers, however, the consequence is satisfaction for a job well done, and preparation for yet another struggle to get it right the next time.

In sum, organizing society absolutely and universally for the production of surplus value is more than an arbitrary, temporary, or reformable condition of life. Rather, the extraction of surplus value is the fundamental law of capital that must extend itself into every fiber of world society until every human being, every human experience, every human act, and every human desire has been commodified and brought into the social relation of capital. Toward this end, the dynamic forms of relative surplus value are put into motion through the creation of new technologies, new class formations, and new governmental regulations. It is this relative form of producing surplus value that the field of the political economy of communication examines, but always within the framework of the production of absolute surplus value. This model is the one that I propose we adopt in our critical work in computers and composition.

THE POLITICAL ECONOMY OF COMMUNICATION

Mosco notes that as a field the political economy of communication developed in response to a profound change in how capital produced surplus value at the macro level—the shift from primarily competitive to primarily corporate capitalism. This shift began in the late nineteenth century, rapidly expanded in the years prior to World War I, and resulted in increased corporate control of the content and technologies of mass media. Scholars began to investigate the social effects of the corporatization of mass communication and have identified four central areas of investigation: the causes of the shift from individually owned to corporately owned communication media; the emergence of a mass consumption economy that accompanied this shift; the entry of the state as the regulatory agent of communication media; and the tendency of western imperialism to impose its culture on colonized people by exporting its communication media (17–21).

If scholars in rhetoric and composition were to develop a corresponding field—the political economy of computers and composition—our starting point would be to identify the macro-systemic shift that has created the urge to incorporate digital technologies into writing and writing instruction. Richard Ohmann's most recent work, *Selling Culture: Magazines, Markets, and Class at the Turn of the Century*, offers us a model of this kind of inquiry in communications studies. This book is a study of the birth of the corporately produced and distributed magazine. Around the 1890s, hundreds of locally owned publications that competed for readership in the capitalist market begin to disappear. Given their dispersed character, these competing publications provided people with relatively heterogeneous interpretations of news, issues, and culture. The market for these locally and individually owned publications, however, was undermined by a new corporatized communications technology: the nationally distributed news, information, and human-interest magazine. In order to analyze this phenomenon, Ohmann first identifies the macro-systemic change that allowed for this sea change in publication practices as the tendency of the European and American economies in the late 1800s to shift from competitive to corporatized and monopolized forms of capitalism, a form Lenin called imperialism. Monopoly capitalism developed, Lenin explains, to consolidate and maximize profit as well as to solve the overproduction crises that ravaged the western economies throughout the latter half of the nineteenth century. Ohmann argues that monopolized and corporatized publications helped to solve the overproduction crisis by creating mass desire for overproduced commodities through new techniques of marketing. A by-product of these new marketing techniques, according to Ohmann, is the appearance of mass culture—the mass production and distribution of a standardized national identity through a nationally homogenized interpretation of cultural and political events.

Following Ohmann's model, the political economy of computers and composition would first have to identify why digital technologies have become an object of inquiry in composition studies. Political economists have already identified that phenomenon as yet another shift in how capital produces surplus value—the shift to globalized accumulation. In this shift, capital attempts to solve its profit crisis by lifting international restrictions on trade. Communications scholars currently analyze the social effects of new, globalized communications technologies. They argue that these new technologies primarily function to extend the social relation of capital by mobilizing previously untouched facets of the cultural life of the world's people into the production of surplus value. They warn that instead of democratizing socioeconomic relations, the new globalized technologies enable more social division through this mobilization.

If the political economy of computers and composition were to follow the communications model, we would also take globalization as our starting point. We would explain the emergence of digital technology in writing instruction as a relative micro-practice of the macro-systemic shift to globalized forms of accumulating surplus value. The work in computers and composition that already takes economics and history into account, however, still lacks a thorough critique of globalization as its starting point, and, thus, makes claims about democratic possibilities that are fundamentally illusory. In what follows, I critique texts by Anson, Berlin, Downing, Moran, and Selfe in order to argue for the necessity of this starting point so that our pedagogical and political activity resists becoming imbricated in globalization. I posit that in the contradictory tendencies of digital technologies to democratize or centralize micro-practices, centralization is the predominant aspect, and this is so because within the social relation of capital, technologies are primarily developed, funded, produced, and distributed to serve the drive for absolute surplus value.[5] Accordingly, the kinds of social practices and relationships that emerge in the social use of these technologies are also primarily determined by capital's drive for absolute surplus value.

CITIZEN-WORKERS, DEMOCRATIC SUBJECTS, AND DEMOCRACY HOPE

In his thinking about the role of composition studies in the twenty-first century, Berlin asks compositionists to embrace, albeit critically, new technologies. Following David Harvey's *The Condition of Postmodernity*, he argues that inflexible regimes of work associated with Ford's assembly line method of industrial production have given way to flexible regimes of work. In the Fordist period, commodities were produced from beginning to end in individual factories. In the post-Fordist framework centralized manufacturing yields to decentralized commodity production: parts of a single commodity are manufactured in various parts of the world and are only brought physically together at the moment of final assembly.

Berlin argues that the present post-Fordist economy creates new regimes of work that require "new forms of cooperation in production, distribution, exchange, and consumption that encourage democratic arrangements throughout the workplace" (224). He understands that these work regimes function dialectically: on the one hand, they enable methods of work that serve capitalist exploitation; on the other hand, these methods can also be turned into their opposite and act to restrict capitalist exploitation. Therefore, in order to assure that democratic interpretations are inscribed in these new methods of work, intellectuals who work within their various disciplines and fields must develop a college curriculum that not only prepares students for work, but does this "within a comprehensive range of democratic educational concerns." In such a curriculum, "Students must learn to locate the beneficiaries and the victims of knowledge, exerting their rights as citizens in a democracy to criticize freely those in power" (223). For Berlin, a new material condition has come into being as a result of new technologies—specifically, a new regime of work that moves the control of decision making from capitalists (who stand above the point of production) to workers who are situated at the site of production. According to this argument, because workers and work sites are dispersed, workers are more able to see their work and discuss their work as it affects human lives. In Marxist terms, workers would become less alienated because their work is more visibly tied to its social purpose. The displacement of decision making from corporate concerns for profit to worker concerns for human lives, therefore, provides the material basis for a democratizing of workers' consciousness.

Many scholars from the field of the political economy of communication, such as Peter Meiksins, have considered the kind of claims that Berlin makes. According to Meiksins, some political progressives claim that "we have entered an age of 'flexible specialization' in which new technologies support loose networks of autonomous producers and create a workplace populated by autonomous, skilled workers" (152). However, Meiksins contends that most progressives see that new technologies "extend employer control over workers, even over long distances, and . . . create automatic systems that can replace the judgement and discretion of expert employees" (152). Meiksins agrees that both aspects of the contradiction do exist and are in struggle, but the working out of this contradiction is influenced by other contradictions of capitalist production. In his view, the contradiction between centralized and dispersed decision-making practices exists within the overarching "conservative character of capitalism, its tendency to frustrate progressive social change" (152). Following Marx's placement of technology as an always relative formation that is primarily subordinant to the formation of absolute surplus value, Meiksins concludes that the "capitalist relations of production constitute technology as a way of controlling and replacing labor, and impede the possibility that technology might democratize the workplace" (158).

To make this case, Meiksins offers a number of studies of new technologies and new dispersed work sites that place decision making in the hands of the workers or managers themselves. In each study, Meiksins notes the same trend. In the dialectical struggle between dispersed and centralized forms of decision making, the need to centralize overwhelmingly dominates and disciplines the dispersed practices by bringing them back into the central domain. One example, at the managerial level, is Silicon Valley. It is often claimed that in Silicon Valley new cooperative relationships among companies and their suppliers are breaking down traditional corporate culture. Information sharing, aided by electronically mediated information exchange, is replacing competition between firms. However, as Bennett Harrison argues in *Lean and Mean*, it becomes evident that over time smaller supplier firms tend to be subordinated by the larger companies they supply. Meiksins explains that "cooperative networks of firms . . . turn out to be vulnerable to the emergence of dominant firms or to intrusion by giant corporations from outside" because the dominant need for the efficient production of surplus value always determines why these practices are developed and maintained in the first place (155). Control from the top reasserts itself because decisions about profitability and productivity supersede any other decision. New technologies must serve profitability; democratic decision making at the managerial level, therefore, exists only so long as it works toward this fundamental goal. As soon as it interferes, the democratized practices are rescinded.

In contrast, among highly paid workers (such as machine operators in the machine tool industry), capital uses new technologies very obviously and unapologetically to deskill these workers and make them redundant for the sake of productivity and profitability. One of Meiksins' studies is particularly salient here. He analyzes the introduction and repercussions of the CNC (computerized numerical control) lathe in the machine tool industry. Equipped with computerized feedback loops, CNCs are able to map electronically and remember the adjustments that workers make in their movements. This mapping allows their skill and knowledge of how to manipulate the lathe to be programmed into the machine, thus eventually making the workers redundant. Once their knowledge is possessed by the machine, the workers are no longer considered to be the creators of that knowledge. In the machine tool industry of the late 1970s and 1980s, the companies using CNC technology promised to move the workers who were made redundant by their own knowledge into even higher levels of skilled work as programmers. However, as Meiksins reports, this promise has not been realized in a corporate culture that is habituated to using machines to save on labor costs, remove troublesome workers, and separate mental from manual labor. Meiksins concludes that the presence of new technologies and the new regimes of work they enable will not change the way that capitalism is organized "as a mode of production based on exploitation and control of labor" (163).

This brief review of Meiksins' research counters Berlin's hope that a new democratic culture could potentially arise from the new work regimes of post-Fordism. As Meiksins suggests, the problem with Berlin's view is that he focuses on the emergence of one new, and very contradictory, micro-practice of relative surplus value: the seeming expansion of cooperative decision making at the managerial level. He isolates this one micro-practice of capitalism from its dialectical relationship with the macro-system: the absolute drive of capital to create surplus value through the creation of proletarianized people. In so doing, Berlin can paint a picture of new democratic possibilities while ignoring the class antagonism between the proletariat and the capitalist class that absolute surplus value requires. Meiksins recognizes that changes in the various forms of relative surplus value do not mitigate the effects of capital's absolute drive for surplus value that require the exploitation of labor. However, he only examines the ways exploitation occurs in the work practices of middle and lower managers and those sections of the working class who are paid above the level of subsistence. In this sense, his refutation of claims about the democratizing effects of post-Fordism does not take into account exploitation as a process of proletarianization. Nevertheless, Meiksins illustrates that in the era of globalized capital, those who receive the brunt of the exploitation may shift, but the exploitative nature of capital continues to be a social fact. When class antagonism is completely unaddressed, as with Berlin, the ideology of democracy hope takes over, diverting our attention from those most exploited by capital.

Like Berlin, David Downing, a proponent of neo-pragmatist pedagogy, also sees the culture of the classroom as the site where new technologies can be used to promote a new democratic consciousness. Downing, however, following John Dewey, is more wedded to the view that new democratic forms emerge out of the new technologies themselves rather than out of the new practices of work these technologies enable, as Berlin suggests. Although Dewey's democratic-socialist political and pedagogical vision failed, Downing attributes Dewey's failure to the failure of the machines on which the philosopher based his vision. Downing argues,

> The electrical/telegraphic print environment . . . still fostered individualism and hierarchy rather than cooperation and collaboration. There was, of course, no way for Dewey in 1894 to predict the potential of cyberspace and virtual reality environments made possible by fiber-optic and micro-chip technology to so alter the classroom and the media in many ways that would be compatible with his own social and political beliefs in collectivity and participation. (186–87)

In Downing's view, Dewey's plan to achieve a truly democratic society failed because the kind of technology required to carry out the plan had not yet been invented. He assumes that "we are going through a cultural revolution in the shift from print to electronic environments as great and significant as the shift from oral to literature cultures 2000 years ago" (193). While Downing notes that the shift in technology is taking place, he fails to situate it within the shift from corporatized imperialism to globalized imperialism.

Ken Hirschkop's work in the field of the political economy of communication cautions against the kind of democracy hope at the heart of Downing's pedagogy. In "Democracy and the New Technologies," Hirschkop analyzes two claims that see democratic possibilities as ultimately depending on the new transmission and information potentials of cybertechnologies themselves. The anarchist view emphasizes the transmission potential and is captured in the identity of the hacker who claims that electronic transmission technologies are inherently slippery forms that, in the hands of skilled technicians, can be used to penetrate and thus to open up systems that were originally designed as closed systems for "military command and control . . . that could survive a thermonuclear attack" (212).[6] Downing's view, however, does not belong to this anarchist/hacker perspective. It belongs instead to the second view analyzed by Hirschkop: the liberal democratic view that is captured in the identity of the electronic referenda democrat. According to this view, the general populace for the first time has limitless access to knowledge and the ability to transmit it; therefore, people can organize this knowledge in order to produce popular power in opposition to corporate power. Only the new cybertechnologies make this possible; thus, democracy is able to function for the first time. "The clear implication," Hirschkop argues, "is that if all could have roughly equal access to these new resources, then the consequences would indeed be democratic . . . [A] better informed citizenry could, given the communicative possibilities of the Internet, wield power through more frequent and more thorough forms of consultation" (214).

Downing articulates this view when he argues that in a print environment, critical exchange is forced to follow a temporally and linearly bound process;

however, new telecommunications technologies allow that process for the first time to be "presented in dialogical form and made available to other researchers without the time lag required of print processes" (198). Instant access to information, therefore, enables democracy; new technology, for the first time, gives everyone this access. According to Hirschkop, both the anarchist/hacker view and the democratic/electronic referenda view are fundamentally joined in the belief that unlimited access to information and the ability to transmit it equals political power. Hirschkop, however, argues that in the history of the democratic struggle, access to information has never been the problem. Therefore, any technology that extends access to information cannot provide the material basis for democratic development. He explains:

> The fundamental inequity of political power does not rest on inequality of information: those who rule do not rule because they know more, . . . but rule whether they know what they are doing or not. Capitalist corporations may well feed information to political representatives and bureaucrats, but their ability to influence them does not depend "in the last instance" on the quality of the information, but on their ability to give or withhold support for state projects or for particular political groups. Private business institutions and their quasi-public associates have financial power and management structures; the state has juridical institutions and the ability to wield coercive force when necessary. (214–15)

The problem, according to Hirschkop, lies in the logic of capital itself: "the imperative of profit" (217). If our information does, on occasion, denude capital of its cover, some change in the relative social relations and forms of production may be adjusted; however, the absolute social relation of surplus value is not adjusted—nor can it be. The class antagonisms inherent in this social relation continue to be reproduced.

Heather Menzies' critique of post-Fordist theory also challenges its claims about the democratizing effect of capital's most recent commodity: information. However, she also points out a more pernicious effect of post-Fordist theory: the erasure of the proletariat as the class forced to live at the level of subsistence. In her study of new work regimes in Canada, *Whose Brave New World? The Information Highway and the New Economy*, Menzies presents a compelling ethnographic investigation of how work practices designed for the production and distribution of information in the globalized economy still require thoroughly proletarianized people. Like Meiksins, she argues that "the ideology of monopoly capitalism" continues to drive the new practices. But her study also identifies the growing divide between the middle class and the proletariat. She defines this divide as that between "the overworked rich with a host of powerful information tools at their disposal" and "the barely working poor on the other side of this income and digital divide" who work "as the hands and voice box of intelligent systems which dictate everything about the job to be done, and monitor every aspect of its performance," such as telemarketers, one of the fastest growing sections of the workforce ("Challenging" 95). In Menzies' view, the new, flexible post-Fordist worker transcends old class boundaries; this new worker can be an upwardly mobile technician or a manager of information systems or a non-industrial proletarian who helps to produce, serve, distribute, and market information as a commodity in the globalized economy. Unfortunately, Menzies' study does not analyze the continued existence of proletarianized industrial workers and the effects of globalization on them.

But recent social movements against overseas sweatshops and against the role of state institutions (such as the International Monetary Fund and the World Trade Organization) in economic deprivation have helped to raise awareness of the continued existence and, indeed, expansion of the industrial proletariat, whose conditions of life and work are straight out of *Capital*. Radical activist groups have continued to pay attention to the existence of proletarianized people, and leftist newspapers and magazines often publish investigative reports on their lives. As an example of this investigative work, I'd like to summarize a well-documented, two-part report from one of these newspapers, *The Revolutionary Worker*, on the living and working conditions of the industrial proletariat in Silicon Valley. The writers show an awareness of post-Fordist theory; however, they apply these new theories without setting aside Marx's claim that capital necessarily rests on the exploitation of laboring people whose lives are valued according to the absolute minimum required for them to subsist.

The article begins with the testimony of Lani Hironaka, the Executive Director of the Santa Clara County Center for Occupational Safety and Health (SCCCOSH), before the California State Senate. Hironaka, playing the same role as the factory inspectors in nineteenth-century England, identifies "overcompetitive subcontracting" as the primary employment practice of Silicon Valley's electronic assembly plants ("Living"). For Silicon Valley's workers directly employed in the electronics industry, this employment practice means "poverty-level wages, piece-rate compensation, chemical and ergonomic hazards, routine health and safety violations, no medical benefits, retaliation, and an immigrant, largely female, non-union work force." She concludes that these elements comprise "what the public commonly refers to as sweatshop conditions" ("Living"). The article notes that subcontracting—a practice characteristic of the flexible, post-Fordist service industry—has also been adopted in industrial assembly plants where commodities are produced for sale in the traditional market. This flexible practice is euphemistically referred to as "outsourcing" and is explained by economists as the normal function of economy to "find" cheaper labor. But the word "find" implies that this cheaper labor is already there. When looked at more carefully, outsourcing or subcontracting functions to create cheap labor, and Silicon Valley serves as a prototype of this process.

The article in the *Revolutionary Worker* goes on to report that in addition to the 100,000 industrial proletarians who work in these sweatshops, an additional 200,000 non-industrial proletarians also work either as service workers in the related industries that electronics requires or as servants to the professional and managerial middle class that the electronics industry has spawned. All 300,000 of these proletarians live in the condition that Marx refers to as immiseration.

Given the cost of living in Silicon Valley, even workers earning $50,000 a year can't afford to rent or purchase property in which to live ("Living"). This lower section of the middle class, in other words, finds it difficult to live at the minimum level to which that class is accustomed. Proletarians, who earn much less than these middle-class workers, commonly rent a corner of someone's living room floor at the rate of $150 to $200 per month in exchange for eight hours of daily sleep time. Many are homeless and sleep on city buses. The article claims that luckier proletarians stuff up to twenty or more members of multiple families into rented single-family homes, or they rent exorbitantly priced garages. For example, a Latina mother of three who works as a night-shift janitor for a large, electronics corporation pays $750 per month (out of her salary of $954 per month) to rent a garage. During the day, when she should be sleeping, she works at the Convention Center to provide for the rest of her family's expenses. These conditions of life are described by Marx as immiseration because proletarianized people whose labor has been valued at the minimum subsistence level live below the minimal standards of comfort established in the society. In the case of contemporary U.S. society, comfort means a balance of work, leisure, and sleep time; a place to live that costs no more than twenty-five percent of a worker's wages with an adequate amount of space for some level of privacy; and a sense of security in knowing where the next meal is coming from, how family members will be clothed, how health problems will be dealt with, and how the family will transport itself routinely to work, school, stores, and so on. When the cost of labor is set at the minimum, the everyday life of those maintained at this level cannot conform to even a minimum standard of comfort. The state, of course, assists capital in assuring that large numbers of proletarians will be available to live at subsistence level by establishing the minimum wage—not in the name of assuring their immiseration, but in the name of protecting them from corporate greed.

The article also details the myriad ways that capital withholds from proletarianized labor not only the most advanced techniques of good labor relations but also the most advanced safety technologies. In the area of labor relations, some of the largest and most capitalized U.S. corporations have severed themselves from having any social responsibility for those who make their commodities. Fortune 500 corporations have made good use of changes in labor law, which since the 1970s has sanctioned subcontracting with its socially irresponsible techniques of managing labor, such as paying laborers "by-the-piece," creating home-based work, and eliminating pension and job security. One Manpower employee interviewed by reporters explains that this subcontractor obfuscates its social irresponsibility in the way that it offers its employees the option of medical insurance. The cost of this insurance is $400 per month, a cost that only managerial personnel who are paid above the cost of reproducing life at the minimum level can afford ("Silicon"). According to its public image, however, the corporation seems to be meeting its social responsibilities to all of its employees. Working for essentially unregulated subcontractors also renders the most advanced safety technologies unavailable to the industrial proletariat. Workers are exposed to extremely dangerous working conditions, including toxic chemicals.

For example, the article reports that a recent explosion at MMC Technology's CD-ROM plant in San Jose sent a splash of nitric acid into the air, and only a trace of this acid needs to be ingested to cause people to "literally vomit [their] guts out" ("Living").

Finally, the article illustrates that labor laws against discrimination have had no impact on the corporate practice of subcontracting. In Silicon Valley, over eighty percent of the industrial proletariat is comprised of immigrant women from over thirty different countries. Karen Hossfeld of San Francisco State University, who has been studying the lives and working conditions in Silicon Valley for twenty years, found that human resource managers are extremely frank about why this is the case. One told her: "I have a very simple formula for hiring ... small, foreign and female.... These little foreign gals are grateful to be hired—very, very grateful—no matter what" ("Living"). Hossfeld was told by another manager that "they won't hire Black people under any condition" ("Living"). Hossfeld's research reveals that ethnic discrimination has as much to do with employers' perceptions of who is least likely to resist as with their racist notions of natural superiority. In this case, African Americans are being excluded, even from the worst jobs, because of their long history of resistance.

The working and living conditions described in the report correspond point-for-point to Marx and Engels' definition of the industrial proletariat. Not all workers experience proletarianization, but capital, as a social relation, requires proletarianization as its foundation. The proletariat sets the standard by which the cost of workers paid above the level of subsistence is gauged. When compositionists talk about the new work practices made possible by new technology, we must not forget that the foundational practice of proletarianization has not disappeared and is absolutely necessary to make the new technologies possible. Every piece of equipment our universities buy for our technologized writing labs has been touched at some point in the production process by proletarianized labor—a most plentiful, yet extremely invisible form of labor. In an address to the Cato Institute in 1997 on the pivotal role of industrial production in globalization, Alan Greenspan remarked, "A global financial system, of course, is not an end in itself. It is the institutional structure that has been developed over the centuries to facilitate the production of goods and services . . . the real side of economies" (244–45). Although he went on to explain that a "much smaller proportion of the measured real gross domestic product constitutes physical bulk today, than in past generations," the real side of economics, which includes information as its newest commodity, has not yet been overtaken by the financial side of economics.[7] It is also important to remember that at the time of his remarks Greenspan had been steadily raising the interest rate of the Federal Reserve for the expressed purpose of increasing the unemployment rate. This is in keeping with Marx's observation that one of the factors that allows for proletarianization is the presence of a "reserve army of labour" (781–94). Industrial workers, it seems, are still with us, no matter how post-Fordist we get. We cannot erase them from our pedagogy and our politics.

Moran and Selfe attempt to include proletarianized people in their analyses of technologizing composition. They harbor no illusions that new technologies will give rise to a new democratic subject, as suggested by both Berlin and Downing. They argue that technology acts in the context of an unjust economic system, a seriously compromised democratic politics, and some very dangerous ideologies. Each proposes that users of digital technology act as democratic subjects to counter the ways it has increased social inequality. Democratic subjects, therefore, must bring about new economic orders.

The contexts that Moran and Selfe bring into the discourse on computers and composition should not be ignored by anyone. Moran reviews the research of economists Paul Krugman and Lester Thurow that shows that the steady rise in per capita income following World War II came to a standstill in the 1970s, and it has regressed ever since then. Krugman's figures show that since 1979, wealth has been redistributed from the poor to the rich, a process he calls "siphoning." According to Moran, Department of Commerce figures up to the mid-1990s corroborate Krugman's findings. Median income has declined in the last twenty years while, as Thurow's research indicates, "the share of wealth . . . held by the top 1 % of the population was essentially double what it had been in the mid-1970s" (216). Moran points out that the effect of the redistribution of wealth from the poor to the rich renders access to new technologies and the distribution of knowledge about them inherently unequal. In our pedagogies, therefore, the battle for equal access to technology is primary; otherwise, the work we do with technology is complicitous in reproducing social inequality. He urges teachers to resist the corporate hype to constantly invest in new technologies, and he urges researchers to launch studies on the social effects of unequal access and the ways in which teachers, students, and communities are coping with unequal access. He argues that this work must be critically undertaken using a Freirean approach that aims to bring the subjects of the research into the critical process of analysis.

Selfe also recounts how socioeconomic inequality is reproduced through the digital divide. However, unlike Moran, she illuminates the ways in which the politics of globalization is implicated in the reproduction of inequality. No longer embroiled in Cold War politics, the Clinton-Gore administration sought to reverse the slowdown in U.S. manufacturing and productivity and to counter threats to U.S. hegemony in the world market by investing in digital technology. Selfe points out that throughout the 1990s the state has enabled and directed the building of a digitalized infrastructure and the preparation of the workforce needed for a digitalized economy—an effort into which the administration has pulled educators at all levels. As a result, education has suffered. During this period, funding to programs designed to increase the amount of student-teacher contact has been cut and reallocated for the purchase of hardware and software. These expenditures are justified and sold to the public in the name of making educational institutions fit to train students for the new economic order.

Selfe argues that through these expenditures public monies also directly support the expansion of digital industries themselves by supplying them with a sure market in which to dump their goods, regardless of their efficacy. Ideologically, the notion is disappearing that education has any other purpose outside of serving the economy. The Clinton-Gore administration, for the most part, convinced educators to value student contact with technology over student contact with human beings by creating yet another literacy crisis—the technological literacy crisis—that has mobilized educators to restrict all educational goals to only those that will help produce the new technologized workforce. While the administration promised that social leveling will result if only educators would goose-step in this war on technological illiteracy, Selfe notes that "this project is likely to support persistent patterns of economically-based literacy acquisition because citizens of color and those from low socioeconomic backgrounds continue to have less access to high-tech educational opportunities and occupy fewer positions that make multiple uses of technology than do white citizens or those from higher socioeconomic backgrounds" (423). Like Moran, Selfe argues that in our pedagogies, as well as in our political practice, the battle over access to new technologies should be primary. This effort would mean combating the racist overtones that always accompany "literacy crises" with local decision making about which technologies students and communities actually need, while engaging in a critical pedagogy that asks students to understand their relationship to technology—not only as consumers, workers, and users, but in a way that helps them to see "the complex relationships between humans, machines, and the cultural contexts within which the two interact" (432).

Both Moran and Selfe argue that, acting as democratic subjects, we must attempt to narrow the socioeconomic gap by expanding access to technology and the knowledge of how to use it. Both advocate that teachers engage in school and public policy debates over access to technology while also using technology critically in their classrooms. But the macro-systemic capitalist structure disappears from both of their visions. Moran dispenses with the need to acknowledge how capitalism is at work. Although he asks us to address the taboo subject of "the distribution of wealth and of social class" and to acknowledge that "computers are, like other goods and services in our economy, available to those with money, and not available to those without money," our descriptions of these social relations should not be too precise, he contends, because this will leave the audience incapable of taking political action (206).

In contrast, Selfe does not seem to be shy about describing the social relations at work. Her stinging critique of the governmental and techno-corporate sectors reveals their special interests. She argues that the Clinton-Gore administration publicly funded the expansion of the electronics industries at the expense of other industries and they did this outside of the democratic process in order to "jump-start the international effort . . . required to exploit emerging world markets" (426). She also points to the antagonistic relation of capital and labor inherent in the expansion of this industry:

> The economic engine of technology must be fueled by—and produce—not only a continuing supply of individuals who are highly *literate* in terms of technological knowledge, but also a[n] ongoing supply of individuals who fail to acquire technological literacy, those who are termed "*illiterate*" according to the official definition. These latter individuals provide the unskilled, low-paid labor necessary to

sustain the system I have described—their work generates the
surplus labor that must be continually re-invested in capital projects
to produce more sophisticated technologies. (427)

Finally, Selfe does not hedge in implicating the role of the Clinton-Gore administration in using the educational system—again, outside of the democratic process of open and public debate—as a conduit for spreading alarm about the crisis of technological literacy in order to push through its economic restructuring. Yet, even though all of her well-documented evidence points to the identity of governmental and techno-corporate interests that have subverted the democratic process, Selfe evades the conclusion her own evidence so powerfully points to: the state serves capital, not the whole people, and capital benefits the few at the expense of the many.

Selfe stops short of critiquing the capitalist state, and, in effect, mirrors the same well-intentioned work of professionals who fought against child labor and the oppression of women in Marx's day. Reforms of capitalist practices can only reshuffle the burden of producing surplus value from one sector of the people to another. In the case of the present drive of capital, Third World people have primarily taken on that burden, along with expanding numbers of proletarianized workers in the capitalist centers. Capitalism's exploitation of proletarian people to create surplus value and the state's legal sanctioning of this exploitation are absolutes; political decisions, the promotion of one sector of the economy at the expense of another, and the shift in who benefits at what time are always relative and subordinate to the fundamental drive for profit. Acting as democratic subjects to struggle for more access to technology on behalf of those currently excluded ignores the fact that all sectors of the world's people are already (or are in the process of) being mobilized into the work of producing surplus value for capital through the full-scale wiring of the world.[8] *How* they are mobilized to do this depends on their class location. Moran's and Selfe's calls for equal access, though well-intended, will result in little more than a shift in the composition of the middle class. Unfortunately, in a capitalist regime, any economic gain made by one sector of the people must be paid for by the increased exploitation of other sectors. Conceptualizing democracy solely as a political category is at the heart of what I call democracy hope. It is inherently problematic in its underlying assumption that social antagonisms are always temporary and relative to specific questions and issues rather than permanent formations of the capitalist structure. When economic relationships are fused with the political, however, it becomes clear that certain social antagonisms cannot be resolved through the political process alone because there is an ethical divide embedded in the economy that cannot be bridged. Although Moran has argued that making naked assertions about economic exploitation paralyzes people politically, I argue against that premise and suggest that we refashion our discourse to include expressions of ethical antagonisms.

REFASHIONING OUR DISCOURSE

Anson's critique of computers and composition demonstrates what an articulation of ethical antagonism looks like. He situates the technologizing of composition in the larger project of corporatizing the university. When we engage in public discourse about the various phenomena that claim our attention, he suggests, we must understand that we are entering a terrain in which one set of values is pitted against another. His argument is constructed around a series of antagonisms and the oppositional ethics behind these antagonisms. He argues that new technologies have primarily been imposed on humanities faculties rather than introduced by us, and he asks us to reject the logic of efficiency that lurks behind the drive to wire everything. University administrators are allocating more and more funding for the development of a wired infrastructure in response to "economic, occupational, and technological" pressures external to the university (262). These pressures stem from "the overriding goal of creating economic efficiencies" to generate "increased revenues" for the university, and some of the actions we are asked to take "frequently clash with some of our basic beliefs about the nature of classroom instruction" (263). Thus, institutional goals are increasingly corresponding to the corporate goal of efficient production. The admittedly inefficient practices that have accompanied the university's pedagogical mission—such as one-on-one interaction between students and teachers in a shared physical space—are being replaced by distance education; standardized Web classes; cut-and-paste reading, writing, and research practices; and standardized multimedia lectures, lessons, and assessment. I would add that through these practices college graduates, as commodities, can be produced more efficiently, and they can be prepared ideologically for the new flexible workforce.

Anson identifies an irreconcilable contradiction between the inefficient pedagogical practices of the traditional university (with its ethics of dialogic discussion and reflection) and the efficient practices of the corporatized university with its delimiting of ethics to those serving the civil sphere only. Anson is not making a sweeping statement against the use of new technology in the classroom. He argues, however, that our pedagogy must not be subsumed by the larger economic drive for efficiency; rather, we must "take control of these technologies, using them in effective ways and not, in the urge for ever-cheaper instruction, substitut[e] them for those contexts and methods that we hold to be essential for learning to write" (263). His language reflects a social antagonism at work; the larger economic drive threatens to "take control" of what our better judgment tells us to do. He does not imply that it is impossible for teachers to act as agents and to resist the way our classrooms are being subsumed by corporate ethics, but he most definitely implies that unless we consciously understand that the technologies are overwhelmingly implicated in the corporate ethic, our work with technologies will be determined by that ethic. His argument, as I read it, is that teachers should identify themselves as antagonistic to the corporate ethic. An identification of this sort produces teachers who will not compromise their ethics when corporations or corporatized administrators try to impose their agendas on them; rather, they will resist those agendas and attempt to politicize others in the process. In this way, a political rhetoric that articulates the ethical antagonisms at work in the economy does not (as it is often assumed in the traditional rules of rhetoric) shut down the possibility of further argumentation, praxis, or consensus building. It does, however, elucidate the ethical divide between competing socio-political interest groups who continue to act even while vying for power.

Anson also warns us that when we uncritically embrace the technologizing of the university as a whole, we are complying with the unethical labor practices that are overtaking our universities (and, I would add, all our institutions). He notes that the drive for efficiency "may lead to even greater exploitation in the area of writing instruction" (263). New technologies have enabled administrators to replace faculty lines with proletarianized adjuncts who are pressured to deskill their work by standardizing instructional material and methods. The use of adjunct labor also allows the university to replicate the highly proletarianized labor practices of piecework and homework—practices that alienate the body and mind of the teacher/worker from students by placing the teacher on a television screen or converting the teacher into the disembodied voice of e-mail.[9] The drive for efficiency results in the increased alienation of the teacher/worker or the student/worker from the product of his or her work. Anson reminds us that these alienating practices are not empowering workers, nor are they breaking down class, gender, and racial distinctions. I would add that the lives of many if not most of the temporary, contract labor that the university employs can be easily characterized as immiserated in that the value of their labor has been set at the level of subsistence. Anson calls for specific political action to restore the educational mission of the university by demanding that the interests of business and industry not be privileged over the scholarly pursuits of teachers and students. He encourages academics to conduct critical research on the impact of the drive for efficiency on teaching and learning. His argument identifies a class antagonism that can only be resolved by reversing the corporate labor practices of the university. He strongly implies that unless we pay attention to the class antagonisms involved in the labor practices of the university, we cannot restore the educational mission of the university or the academic freedom it requires.

The antagonistic ethics attached to this class antagonism is clearly exemplified in the eruption of the movement of adjuncts and graduate students to unionize.[10] But even though political struggle that directly addresses class antagonism has broken out on campuses, the class antagonism embedded in issues surrounding technology has, for the most part, been ignored. A case in point is the failure of the professoriate to forge a national movement or coalition against the rapid push toward distance education that is being established under the University of Phoenix model. Again, I argue that until the antagonism between corporate and educational ethics enters the discourse, the kinds of compromises over distance education that are being struck between departments and individuals and the highest level of administrators and their boards of regents will go unchallenged. This is a movement that is yet to begin, but it is a movement that, if it does emerge, will directly confront the deployment of technology in the service of the globalization of capital.

Discourse that recognizes the existence of class antagonism and its accompanying ethical opposition has been making its way back into the American political sphere, and it is often planting itself squarely on the front lawns of universities. The Students Against Sweatshops (SAS) movement has often used antagonism as a warrant for its claim that the university must sever it contract with corporations who hide their complicity in the creation of immiseration by outsourcing their labor to local contractors and tyrannical governments while at the same time claiming that globalization is raising everyone's standard of living. When SAS started, the proletarianization that is the fabric, so to speak, of the Nike clothing that is donated to college athletics programs was completely invisible. In a few short years, the class antagonism represented in almost all of the clothes we wear has now been made so visible that Nike, as one of the most targeted corporations, has rescinded its contracts with at least two universities rather than comply with the standards of labor practices that those universities eventually adopted. Nike's action suggests that a proletarianized population is an absolute requirement for the creation of surplus value. Clearly, in order for the capitalists who own Nike to continue living at the standard to which they have become accustomed, the people who make their shoes and jerseys must continue to live and work at the level of subsistence that Nike determines and controls. Class antagonism and its accompanying immiseration is just as embedded in the machines that seem so innocently and neutrally perched on the desks of the computer labs where we teach composition. Before we make claims about what these machines allow *us* to accomplish—claims about nonlinear thinking, reading, and writing and about how those practices have decentered *us* as subjects, somehow freeing *us* from the bonds of the rational world view—and before we make claims that if only more of America's people could have access to these technologies somehow a social leveling would occur, let us first consider not only the presence of the programmer laboring over all those zeros and ones in our machines, but also the presence of the minimum-wage worker who could have died making the CD we just popped into our machine. So long as there is class antagonism, more democracy *for us* is always extracted at the cost of more immiseration *for them*. Also at the heart of our democracy hope is the hope that someday everyone will belong to the middle class. Seeing computers and composition through the lens of political economy will make this hope clearly untenable.

I have argued that in order to speak ethically compositionists who theorize the role of technology in writing instruction must not forget political economy. Electronic technologies, like all technologies of the capitalist era, appear temporarily and relatively only as they continue to serve capital's absolute drive to produce surplus value. The appearance of various classes in this production process is equally temporary and relative to the macro-systemic shifts that have taken place throughout the capitalist era. Only one class, the proletariat, is a permanent feature of capitalism. Globalization, which is increasingly dependent on new technologies, is only the latest macro-systemic shift developed to maintain capital as a social relation, albeit a more internationalized one. This shift has destroyed huge sectors of U.S. industrial workers who were being paid above a mere subsistence level, throwing them back into the commodity market where they have found their labor devalued, while allowing room for other ex-industrial workers to join the reconstituted middle class whose work has become increasingly defined as producers, distributors, and managers of information. But the globalized economy has also created, on a world scale, the largest number of proletarianized people that has ever existed in the history of capitalism. Our practices with new technology are inextricably connected to capital's proletarianization process. The theoretical lens of Marxist political economy does not allow us to divorce political ideals, such as democracy—or, for that matter, socialism—from real economic relationships. Once this connection is made, ethical dividing lines that reflect the class antagonism inherent in capital come into sharp focus.[11]

NOTES

1. Trimbur reminded me of the term *citizen-worker*, which he coined in his review of Berlin's last book.

2. Mailloux has recently posited that rhetorical studies offers the opportunity for the renewal of interdisciplinarity between English and communications studies.

3. The existence of such a gap is so widely accepted that the Department of Commerce has been collecting data to track its movement since 1994 in a series of reports entitled "Falling through the Net: Defining the Digital Divide."

4. In a footnote, Marx explains that he uses the derogatory term "wretch" because it is "the technical expression used in English political economy for the agricultural labourer" (517).

5. I analyze these claims using a dialectical methodology from the Marxist—and, particularly, Maoist—tradition. This methodology understands that the "life" of any phenomenon is characterized by a multiplicity of contradictions. Within any one contradiction, the oppositional aspects never have equal political strength or importance. Rather, one aspect may dominate the other for long periods of time. In addition, among the plethora of social contradictions emerging from capitalism in all of its forms, certain contradictions usually dominate others for long periods of time. While these two precepts of contradiction only begin to broach the body of Marxist dialectical theory, they are adequate for the present analysis.

6. Edwards traces the American origins of digital computation to military ballistics research during World War II, and he analyzes how the military doctrine of "command, control, communications, and information" is congealed in the computer and Internet. Haraway refers to his work in "Manifesto."

7. Schiller offers a compelling analysis of how information came to be produced, distributed, and owned as a commodity.

8. Drawing on the work of Foucault, Robins and Webster conceive of the current communications revolution as a continuation of the profound restructuring of "the micro-systems" of power evident in everyday life since the inception of Fordism. They challenge the idea that post-Fordism represents a break with the power relationships established by Fordism as a micro-practice. They reach this conclusion by "reinstating the concept of totality" to the study of everyday life by situating everyday life within the "historical trajectory of the search for capital accumulation and obstacles placed in the way of this endeavor" (47). They argue that with the onslaught of monopoly capitalism, all of cultural life, not just work life, becomes absolutely mobilized for the production of surplus value.

9. I want to stress here that using Web boards or e-mail is not in and of itself an inherently alienating practice; however, given that new technology is overdetermined by the corporate agenda, the majority of practices with it will be interpellated by that agenda.

10. Nelson and Watt provide ample evidence of how these class antagonisms are constructed within the entire structure of the university.

11. I want to express my appreciation to John Trimbur for his careful and gracious reading of this manuscript, and for the groundbreaking work of James Berlin, without which I doubt my commitment to finding a place for Marx in composition studies would have persisted. I thank my colleagues Jean Kreis and Cathy Chaput for their invaluable criticism at key points in the drafting of this article, and Ken McAllister for all the lunches, coffees, and beers over which we conspired to out-think the machines.

WORKS CITED

Althusser, Louis, and Étienne Balibar. *Reading Capital.* 1968. Trans. Ben Brewster. London: Verso, 1997.

Anson, Chris M. "Distant Voices: Teaching and Writing in a Culture of Technology." *College English* 61 (1999): 261–80.

Berlin, James A. "English Studies, Work, and Politics in the New Economy." *Composition in the Twentieth-First Century: Crisis and Change.* Ed. Lynn Z. Bloom, Donald A. Daiker, and Edward M. White. Carbondale: Southern Illinois UP, 1996. 215–25.

Dinwiddy, J.R. *From Luddism to the First Reform Bill: Reform in England 1810–1832.* London: Blackwell, 1986.

Downing, David B. "The Political Consequences of Pragmatism; or, Cultural Pragmatics for a Cybernetic Revolution." *Rhetoric, Sophistry, Pragmatism.* Ed. Steven Mailloux. Cambridge: Cambridge UP, 1995. 180-205.

Edwards, Paul N. *The Closed World: Computers and the Politics of Discourse in Cold War America.* Cambridge: MIT P, 1996.

Foucault, Michel. "The Discourse on Language." *The Archaeology of Knowledge and the Discourse on Language.* Trans. A.M. Sheridan Smith. New York: Pantheon, 1972. 215-37.

———. "Prison Talk." *Power/Knowledge: Selected Interviews and Other Writings, 1972-1977.* Ed. Colin Gordon. Trans. Colin Gordon et al. New York: Pantheon, 1980. 37-54.

Greenspan, Alan. "The Globalization of Finance." *Cato Journal* 17 (Winter 1998): 243-50.

Haraway, Donna J. "A Manifesto for Cyborgs: Science, Technology, and Socialist Feminism in the 1980s." 1985. *CyberReader.* Ed. Victor J. Vitanza. Boston: Allyn, 1996. 372-412.

Harrison, Bennett. *Lean and Mean: The Changing Landscape of Corporate Power in the Age of Flexibility.* New York: Basic, 1994.

Harvey, David. *The Condition of Postmodernity: An Enquiry into the Origins of Cultural Change.* Oxford: Blackwell, 1989.

Hirschkop, Ken. "Democracy and the New Technologies." McChesney et al. 207-17.

Lenin, Vladimir I. *Imperialism, The Highest Stage of Capitalism: A Popular Outline.* 1917. New York: International, 1939.

"Living on the Bottom of Silicon Valley: Proletarians in California's High-Tech Zone." *Revolutionary Worker* 1054 (14 May 2000): http://www.rwor.org/a/v22/1052-059/1054/silicon.htm (15 Dec. 2000).

Mailloux, Steven. "Disciplinary Identities: On the Rhetorical Paths Between English and Communications Studies." *Rhetoric Society Quarterly* 30.2 (2000): 5-29.

Marx, Karl. *Capital Volume One.* 1867. Trans. Ben Fowkes. New York: Vintage, 1976.

Marx, Karl, and Frederick Engels. *The Communist Manifesto.* 1848. New York: Pathfinder, 1987.

McAllister, Ken S. "Care or Cutting Edge? A Review of Three Books about Computer-Enhanced Pedagogy." *Rhetoric Review* 18 (1999): 192-99.

McChesney, Robert W. "The Political Economy of Global Communication." McChesney et al. 1-26.

McChesney, Robert W., Ellen Meiksins Wood, and John Bellamy Foster, eds. *Capitalism and the Information Age: The Political Economy of the Global Communication Revolution.* New York: Monthly Review, 1998.

Meiksins, Peter. "Work, New Technology, and Capitalism." McChesney et al. 151-64.

Menzies, Heather. "Challenging Capitalism in Cyberspace: The Information Highway, the Postindustrial Economy, and People." McChesney et al. 87-98.

———. *Whose Brave New World? The Information Highway and the New Economy.* Toronto: Between The Lines, 1996.

Moran, Charles. "*Access*: The 'A' Word in Technology Studies." *Passions, Pedagogies, and 21st Century Technologies.* Ed. Gail E. Hawisher and Cynthia L. Selfe. Logan: Utah State UP, 1999. 205-20.

Mosco, Vincent. *The Political Economy of Communication: Rethinking and Renewal.* London: Sage, 1996.

Nelson, Cary, and Stephen Watt. *Academic Keywords: A Devil's Dictionary for Higher Education.* New York: Routledge, 1999.

Ohmann, Richard. *Selling Culture: Magazines, Markets, and Class at the Turn of the Century.* London: Verso, 1996.

Peel, Frank. *The Risings of the Luddites, Chartists and Plug-Drawers.* 1880. London: Frank, 1968.

Robins, Kevin, and Frank Webster. "Cybernetic Capitalism: Information, Technology, Everyday Life." *The Political Economy of Information.* Ed. Vincent Mosco and Janet Wasko. Madison: U of Wisconsin P, 1988. 44-75.

Sale, Kirkpatrick. *Rebels Against the Future: The Luddites and Their War on the Industrial Revolution.* New York: Addison, 1995.

Schiller, Dan. "How to Think about Information." *The Political Economy of Information.* Ed. Vincent Mosco and Janet Wasko. Madison: U of Wisconsin P, 1988. 27-43.

Selfe, Cynthia L. "Technology and Literacy: A Story about the Perils of Not Paying Attention." *College Composition and Communication* 50 (1999): 411-36.

"Silicon Nightmares: What It's Like to Work in the High-Tech Sweatshops of Silicon Valley." *Revolutionary Worker* 1055 (21 May 2000): http://www.rwor.org/a/v22/1052-059/1055/silic.htm. (15 Dec. 2000).

Smith, Adam. *An Inquiry into the Nature and Causes of the Wealth of Nations.* 1776. Chicago: U of Chicago P, 1976.

Trimbur, John. "Berlin's Citizen and First World Rhetoric." *JAC* 17 (1997): 500-02.

U.S. Department of Commerce National Telecommunications and Information Administration. Reports, Filings, and Related Material. http://www.ntia.doc.gov/reports.html (19 June 2000).

6
CIRCUITOUS SUBJECTS IN THEIR TIME MAPS

James J. Sosnoski
Ken S. McAllister

Over two decades ago, Fen Labalme designed the prototype of a postmodern newspaper that he was then designing as part of a media and technology project at MIT. Stewart Brand, the Director of MIT's Media Lab, summarized Labalme's project in his history of the Media Lab:

> *NewsPeek*, a selective home-publishable semiautomatic electronic newspaper that knows the reader, [is] made of material drawn daily from Dow Jones News Retrieval, Nexis, XPress, and wire services, along with television news. Walter [Bender, the director of the project] punches it up on his monitor screen. Topic headlines in different colors indicate "international," "Technical," "Financial," "Mail," "People," etc. When he slides his finger across the screen, the image on the screen slides with him, revealing more text. He runs his finger across a lead paragraph, and that story fills the screen. He calls for other newsclips on the topic, and three come up, one of them colored pale yellow like aging newsprint, indicating it's an old item.
>
> Illustrations on the screen in color, such as the map of Cuba or the photograph of the President, are drawn locally from a videodisc capable of holding 54,000 such images, the sort of thing that might be mailed out monthly by a subscription service. When Walter touches an article under "Today," suddenly the illustration comes to life, flames and smoke pouring up, a television voice announcing, "In Mount Bellevue, Texas, today there was an explosion at an oil refinery that set off a spectacular fire. Flames from burning

> propane, butane, and gasoline towered 800 feet...." The clip was
> captured from the evening news by *NewsPeek* and formatted into
> the presentation. The most significant item on *NewsPeek*'s front
> page ... [is] the user's own electronic mailbox.... "It's news only to
> him, but it's the most important of all." (*Media Lab* 37)

Back in 1981, when ideas like this were just beginning to accrue popular attention, Labalme's *NewsPeek* seemed a futuristic vision. Now, it reads like a description of familiar communication tools such as *Opera*, *Netscape*, *Internet Explorer*, and *Firefox*, as well as a rather simplistic preview of graphical and tactile user interfaces, the likes of which have now been iconized in movies like *Minority Report* and *The Matrix Reloaded*. Today, newspapers, TV networks, and political groups all have web sites that more or less correspond to experiments done during the eighties at MIT's media lab because in the early twenty-first century such "experiments" have become commercially viable applications. As a consequence, the critique that is usually reserved for the prophetic and naively enthusiastic utterances of the likes of Labalme and Brand have been quieted by the somewhat embarrassing fact that the utterances have not only become manifest but popular.

As readers of *JAC* are well aware, a similar phenomenon has occurred in the disciplines of English and communications studies. The interface between text and cultural study has now been sufficiently technologized and transformed that scholars in these and related fields might usefully refer to the various multimedia applications in the humanities not as "NewsPeeks" but as "CultPeeks." Students can now access hundreds of hypermedia resources via the web, CD-ROM, and DVD that allow them to become immersed (in varying degrees) in the intricacies of a specific text's production, from Tennyson's poetry and performances of *Faust*, to interpretations of Shakespeare in cinema and interactive digital media. Several MOOs facilitate the study of canonical literature and ancient Greek and Latin authors, and there are at least two virtual reality renderings of important sites in the Mediterranean Basin designed to be explored by classics and art history students. Students of rhetoric, on the other hand, can go to LinguaMOO, where they can listen to rhetors in its theatre while eating virtual popcorn. What seemed so futuristic in 1989 is now commonplace. Many classrooms are now beginning to approximate fully immersive electronic educational environments (or as we shall refer to them throughout this paper, EEEs), because they primarily offer students and teachers just a physical space wherein they may login to a multitude of virtual spaces. By "electronic educational environment," we simply mean an interactive learning situation in which various media are programmatically interfaced as a massive dynamic hypertextual database that can be drawn on to provide specific courses of study. We also concur with the findings of the Committee on Virtual Reality Research and Development that "the purpose of a virtual environment system is to alter the state of the human operator or the computer" (Durlach and Mavor 2).[1] Today's computer-enhanced classrooms certainly fit these descriptions, connected as they often are to numerous online indices, search engines, and archives;

loaded with a plethora of software to edit photos, illustrations, movies, and audio files; and enabling communication with other computer users around the world using anything from text-only email to streaming video connections. Such technologies aren't restricted to the computer-enhanced classroom, of course. Many students now have access to these kinds of hardware and software technologies in their home, making their bedrooms and dorms into EEEs in their own right.

But not all EEEs are alike. In this essay, we distinguish between two types of EEEs: *non-dialogical*, in which users navigate in solitude and that are principally published on CD-ROMs, DVDs, and static websites; and *dialogical*, in which users necessarily encounter others in such a way that their exchanges become part of the learning process. Dialogical EEEs usually take the form of MUDs, MOOs, chat rooms, web forums, and, at the leading edge, networked 3D environments similar to (or modifications of) massively multiplayer online games (MMOGs).

That interactive multimedia resources like *NewsPeek* grow increasingly familiar makes what happens to us as we use them also seem commonplace. This is not an unmitigated blessing because what appears to be normal does not appear to require scrutiny and thus is sometimes accepted without reservation. In our view, advocates of EEEs (among whom we count ourselves) need simultaneously to be critical of the emerging pedagogies such technologies spawn. Hence, in the following reflections, we link critical commentaries on the formation of subjects offered about print culture with the rapidly growing educational technoculture sprouting up in departments housing various forms of cultural and media studies.

Though humanities scholars have learned to theorize interpellation, we are often unaware of how subjected our subjectivities can become. The attention we pay to the circumstance that we constitute printed texts from the intellectual frameworks we bring to them should not be allowed to obscure the circumstance that, in doing so, we frame meanings out of our memories of prior readings of writings we did not author. In the highly subjective process of reading either print or electronic matter, we write ourselves largely from interpretations made available to us by others. In reading, we subjugate ourselves in the sense that the very subjectivities that frame our reading are themselves altered by what we read. Writings are not merely subject matters we read, they are subjugations (instances of acculturation) we often unconsciously accept. These considerations may seem unimportant in considering cyberspace if we disregard the circumstance that its *texture* is discursive and has to be read. In what follows, we call attention to two forms of subjected subjectivity that we have experienced in the virtual world of cyberspace: circuity and dehistorization. As a point of departure, we ask: are readers of hypertextual, multimedia databases actors who perform scripts written for them?[2]

READING AND WRITING IN CYBERSPACE AS PERFORMANCES OF SYMBOLIC ACTIONS BY VIRTUAL SUBJECTS

The recent propagation of wireless broadband and the latest 3D graphics capabilities of tablet-PCs, cell phones, and PDAs make it reasonable to project that within the next five or so years, students will connect to EEEs so much like *NewsPeek* in their customizability that we can give them the generic name, *CultPeeks*. We use this term because such EEEs, which tend to be non-dialogical, configure users' experiences of the research "motive" built into such educational environments as little more than a "peek" at the complexity of the subjects being studied. Similarly, "cult" configures the microcultures that are spawned by the users' idiosyncratically selective "peeks" at cultures represented by the authors of these EEEs. Like Walter Bender, the director of the *NewsPeek* project at MIT, students—economically privileged ones, at least—will call up such environments on their wireless devices from wherever they can gain access to a global computer network: campuses, public libraries, train and bus stations, airports, and any public building.[3] Subject listings in a U.S. Studies Program, for example, may use different colors to indicate "Early U.S. Politics," "19th-Century U.S. Technoculture," "U.S. Literature from 1900 to the Viet Nam War," and "Contemporary Trends in the U.S. Entertainment Industry." When Walter (on whom we'll turn the table for a moment) slides his fingertip across the screen to "U.S. Literature from 1900 to the Viet Nam War," the screen slides with him, revealing a long list of names and places. He runs his finger across the names to one he does not recognize—Agee, James. Out of curiosity, he touches Agee's name and a list of annotated dates appears. Recognizing a magazine title, Walter picks out "1936, *Fortune* Assignment in Alabama" and a story about Agee's trip with Walker Evans to Alabama, formatted as if it were the magazine itself, takes over the screen. Noting the availability of visuals on the topic, he calls for them. A dozen of Walker Evan's remarkable photographs come up. A close-up of a young woman in a straw hat with strangely sad eyes peers at him; another of a rural post office catches his attention because the "DRINK COCO-COLA" sign oddly overwhelms the designation "U.S. Post Office, Sprott Ala."; a third, of two work shoes sitting in the parched dirt, puzzles him. With his fingertip, he traces a question mark over the image—a technology enabled through the system's gesture recognition software—and a text from *Let Us Now Praise Famous Men* (1939) appears describing the homes of the Alabama sharecroppers in Evans' photographs.

Now fascinated, he searches for information about Agee. A photograph of the writer in a rumpled suit fills the screen; then in the right hand corner a range of biographical topics quietly becomes visible in a window. Noticing "Hollywood," Walter taps it. A list of films shows up in a window in the lower right hand corner of the screen. He touches *The Night of the Hunter* by accident; suddenly the screen comes to life in the figure of Robert Mitchum with "hate" tattooed across the fingers of his left hand and "love" across those of his right. The film freezes, becoming a backdrop for a superimposed menu: "Load Movie ($2.00)" and "Play Game (Free Demo)." Having time for neither (though interested in both), Walter backtracks and identifies a biographer. James Barson, as a fully rendered three-dimensional avatar (built from a generic motion-capture dataset) comes up on screen and says "James Agee's Pulitzer prize winning *A Death in the Family* gives us an evocative view of his childhood in Knoxville Tennessee. . . ." Walter traces the "mute" symbol on the screen and uses another symbol to call for the text of Barson's remarks, which obediently appear in the upper left hand corner of the screen. After skimming the bio's overview, he browses through the headings of Agee's film reviews for *Time* and *The Nation*, eventually settling on an essay on silent comedy, having been attracted by the image on a *Life* magazine cover for September 3, 1949 that had flashed up as a cross-reference in his archive file.

From the perspective of Kenneth Burke's dramatism, users of multimedia systems configured in the CultPeek mode like Walter are performers whose discursive interactions with the database (often recorded as a "history") can be understood as the story (or drama) of reading the database. A subject's time in the virtual world of a database becomes part of the history of that virtual world. These histories are chronologies of choices made along the subject's journey through the database. Of course, many (if not all) of these "choices" have been scripted into the database.

Construing databases as virtual worlds whose histories are comprised of their users' decisions allows us to construe virtual subjects as agents with motives and to inquire into the extent to which becoming a performer in someone else's drama prescribes identification with a set of motives. In this regard, Kenneth Burke's delineation of motives in his various rhetorics is helpful since he shows how discourses that do not appear to be dramas are nonetheless dramatic. From his theoretical perspective, databases can be analyzed as webs of motivation since using them is a symbolic action comprised of agents, acts, motives, and consequences no matter how virtual. Though we rarely do so, we might think of even a simple listserv as a dramatization of motives or of a database as the terrain over which an investigative journey is taken. In either case, histories are made as discourses are built out of chains of decisions.

Though such hi*stories* are minimal—hardly more than chronologies and object sequences—nonetheless, when actions are verbalized, they reveal motives because "performance-verbs play a fundamental part in the description of human actions" (Kenny 183). In his study of the verbalization of action, Anthony Kenny writes,

> Very often, what happens when a human being performs an action may be described as follows. First, there exists a state of affairs of which the agent disapproves; then the agent does something; after his action there exists, in place of the original state of affairs, a different state of affairs of which he approves. . . .
> Wherever this scheme of description of action applies, there will be room for three main types of explanation of action. An action may be explained by reference to the unwanted state of affairs,

> which preceded it, or by reference to the wanted state of affairs which was, or was expected to be, its upshot, or by some form of explanation which alludes to both of these together. (90–91)

Narratologists identify this pattern of description of human behavior as a "story." To put an action into words is to narrate. Part of the force of Kenneth Burke's work is to show us that many forms of discourse (for instance, arguments) can be understood as symbolic actions. A depersonalized argument—a flame in an ICQ chat room, for instance—once returned to its context of debate, quarrel, or controversy, for example, may more easily be discerned as discourse than when it is absent from its context. An examination of listserv logs or web forums as the context for the desires and conflicts expressed therein makes this narratological process more apparent because such logs (records of asynchronous dialogue) are linked to a longer history of readings and writings (arguments) that make up the entire database. Considering that listserv and web forum logs are often quite volatile (in the emotional, not the electronic sense), it is similarly apparent that critical arguments usually begin in desire and conflict: interlocutors want to be right and will become antagonistic to show that they are. From this perspective, arguments can be perceived as the human dramas they invariably are, and can be understood to have histories. As narratologists have claimed, stories often begin with states of desire or conflict that represent the desires and conflicts audiences wish to resolve strategically in terms of their own interests. These interests are often obscure, especially in virtual worlds.

When inhabiting a virtual world, a person en-roles a virtual persona who is also a subject, the "I" of the virtualized discourse. Other subjects online synchronically cannot tell whether a virtual subject is speaking truths or lies, whether the discourse is governed by a male or female, or whether it resembles the user's real-world behavior. Nonetheless, we recognize a virtual subject as a legitimate agent of discourse—that is, as the authoritative "I" who speaks and who hails others through virtual discourse. Though we may know nothing of the person behind the avatar, he or she acts through this actor and occupies a position relative to the other virtual subjects. The avatar's "I" is subjected to the discourse invented for it. In sum, as Kenneth Burke might say, the virtual subject is a symbolic actor whose actions have discursive consequences that are usually experienced as outcomes of a complex of desires and conflicts. We term this complex the *virtual dialectic*.

The outcome of a performance within the virtual dialectic is bound up with a desire that the audience—other virtual agents and their puppeteers—will respond in particular ways to the discursive world presented to it.[4] For this to happen in the virtual world, the audience for a message has to accept *their implicated subject position*. Consider this example: a group of longtime online companions set off on a new adventure in a massive multiplayer online role-playing game (MMORPG); none of the players has ever met one another in real life and the VR world's protocols discourage players from "breaking the spell over the land," a euphemism for talking about life on "Earth" where people have real lives, families, and jobs. As this party sets off, it encounters a wandering healer, all alone. After a brief exchange, the party—short on healers—decides to invite their new acquaintance along, even though the avatar is very inexperienced. As the party travels across the land, making small talk about the advantages of chain mail over plate armor and keeping an eye out for brigands, the new healer makes the inevitable "newbie" error and breaks the spell: "I'm a manager at a Blockbuster Video store in Detroit. What do you all do?"

The healer-avatar's utterance admits the real world into the virtual one, catalyzing a transformative moment. The healer's "I" in the players' innocuous, but consistently en-roled, chit-chat gives the lie to *all* the avatars by hailing the other puppeteers with a "you" situated in a real- (not an MMORPG-) world context. Like many transformative moments that occur when people traverse the increasingly ambiguous boundary between real and virtual dialectic—and the discourses that comprise them—this one is violent. The other players have been forced to abandon their fantasy—if only for a moment—and remember the context of the game they are playing. Their immersion is withdrawn, and they are forced to see again their own real lives, families, and jobs.

Each member of the party has several options for responding to this violent interpellative moment. One might accept the healer's reminder and simply answer the question: "I teach high-school English." Another might accept the reminder but refuse to answer, choosing instead to offer the new player some instruction: "In this world we don't talk about Earth. . . . It defeats the purpose of being here." In such a case, a player acknowledges the truth of the reminder but informs the new player that such reminders are unwelcome; in this VR world, it's rude to offer someone a subject position they can't refuse. Finally, a player might choose to ignore the interpellation, either by not responding or by turning the interpellative moment on its head: "What's a Blockbuster Video store? I've never seen one of those here before." With this latter response, the healer's player is now forced to choose between continuing to occupy a subject position in the real world or to reconstitute the healer-avatar's subject position.

Such moments are common in MMPGs and other dialogical databases, and are negotiated in just the ways described above. Each negotiation carries with it a history—one that not only comprises the past but also constitutes the present and future—that must be en-roled. The player who fails to properly en-role is corrected, and in this way, such environments are self-policing. And once a player concedes to the en-rolement process, her or his avatar's history becomes fixed (though not necessarily forever) according to the terms of the negotiated, which is to say the implicated, subject position. And behind all of these overt interpellative actions in the EEE, there is another one, a shadow interpellator haunting the system: the designer. It is designers who make inter-EEE roles such as "healer" inhabitable in the first place, and while players may self-police the VR world, they rarely police the world-builders themselves.[5] It is at this very point, however—where virtual and real interpellations collide—that virtual and real dialectics coalesce. The desires and antagonisms of the real world invade upon the virtual world and vice versa, transforming the players and their avatars simultaneously.

When dealing with dialogical databases, we may say, then, that the structure of motivation is dramatized: the actors (or at least their avatars) are present, and their desires and conflicts are fairly evident. Though not always obvious, motives in dialogical databases are attributed during the dialogue, and the interpersonal dimension of the inquiry is palpable. The motivational patterns built into non-dialogical databases, however, are far less accessible: the actors are masked by corporate invisibility—far more so even than with the MMPG designers described above—and their motives camouflaged by flamboyant appeals to imagined and shared values of the commonwealth or sometimes to "good old-fashioned commonsense." A Burkean perspective suggests, however, that non-dialogical electronic educational environments are dramatic symbolic actions whose structures of motivation (that is, implicated subject positions) necessarily evade easy detection. In the next section, we describe the circuits of motivation built into the search routines of non-dialogical EEEs. These routines produce what may be called *circuitous subjects*.

CIRCUITOUS SUBJECTS

Electronic educational environments operate through electrical transmission paths—"circuits"—along which a continuous flow of signals form an electronic discourse. Such circuits of signals, like other discursive formations, invite a variety of subject positions. People are positioned by the ways in which they both talk and think about themselves and the ways others talk and think about them. They are placed in positions relative to others in hierarchically structured systems of power. Many readers, for example, are probably teaching assistants, or assistant, associate or full professors. A few may be university deans, provosts, or college presidents. Similarly, in EEEs, users are guests, subscribers, testers, clients, or contributors. These titles represent their bearers' privilege to exercise different degrees of power in higher education and within particular EEEs, respectively. Persons designated "faculty" typically think of themselves as inhabiting a combination of roles—rhetoricians, historians, political economists—which, considered separately from the powerful institutions that support these roles, might well seem subordinate to "faculty." These positions are joined to specific functions in the discursive world faculty inhabit. Such persons are also performers, actors who "teach," "research," and "edit," for example. And notably, these roles do not usually also include positions such as "clerk," "bursar," or "sanitation engineer"; while these roles are neither more nor less important than those a faculty member plays, the social construction of faculty roles has been manipulated such that they retain more cultural capital than those associated with building maintenance and office work.

In Althusser's famous view, we acquire our subject positions by being called to them: "I shall then suggest that ideology ... 'recruits' subjects ... or 'transforms' ... individuals into subjects ... by *interpellation* or hailing, and which can be imagined along the lines of the most commonplace everyday police (or other) hailing: 'Hey, you there!'" (71, 174). An individual so interpellated or hailed recognizes "that the hail was 'really' addressed to him, and that it was 'really' him who was hailed (and not someone else)" (174). An individual, in being called "professor" or "teacher" is hailed by the institution—including all its personnel, even students—in which the terms "professor" and "teacher" function and have power. That individual recognizes that she or he is "really" a professor. She or he thus believes in the reality of professorhood and all that it entails within the institution of the university. According to Althusser, "The existence of ideology and the hailing or interpellation of individuals as subjects are one and the same thing" (175). Persons are subject to discourses that position them in particular mappings of a social world and assigned correlative functions that imply motives designed to justify the system. This is just as much the case in the virtual environments designed by interpellated subjects as it is in the real world that those subjects themselves inhabit.

In educational electronic environments, then, persons are also called to subject positions. Once inside the EEE, where their cultural world is mapped by programmers or other high-level functionaries, they are called upon to make decisions.[6] Because the circuits of decision-making are pre-motivated by their designers and developers, virtual subjects are positioned within the apparatus of the EEE in much the same way that Althusser describes interpellation. The arrangement of data in the database—what programmers call "structure"—only allows for a restricted range of motives. The user who searches the database and makes a list must, therefore, subscribe to at least one of the predetermined motives in order to proceed with him or her work. In order to generate a history of the "game film"—a term coined by Judd Ethan Ruggill to designate movies based on computer games—using IMDb (www.imdb.com), for instance, one must first register at the site, login, learn the database's interface, then master the web application's search engine before beginning to compile such an unusual filmography. Using the wrong terms—terms not implemented by the developers—results in erroneous or empty results. A search for "video games," for example, turns up numerous hits for actor David Games and for director Marcelo Games. In order to find films based on games, a user must experiment until she or he finds a search term and pattern that "makes sense" to the application. Once that rubric is identified, of course, it also seems then to "make sense" to the user. The relationship between video games and films, in other words, is not a motive the program admits, and so the CultPeek that might eventually emerge from such human-computer interaction necessarily implicates the ideologies of both the user and IMDb's developers.

It would be easy enough to take a fatalistic view of the CultPeek scenario just described and conclude that the dramatic transformations brought about by digital telecommunications will only "speed up" the effects of cultural hegemony and posit that we are always already unwitting dupes of the programs we use. We are not quite so fatalistic but in our experience—as both users and developers—electronic educational environments differ significantly from book-oriented learning environments with respect to the construction of intellectuality. These differences are due primarily to the circuitry of virtual subject positions.

The rapid changes in subject positions required by computer programs in general—even basic word processors hail users into particular socially defined roles—and by EEEs especially, form a circuit, giving the impression of coherence. A "circuitous subject" is one who, while online, is routinized through a particular sequence of implicitly motivated performances that resolve back to the starting point, thereby constructing an apparent continuity. In other words, a string of pseudo-random subject positions (understood here as "roles" that prescribe practices within a fixed range of motivation) is given coherence by the task that makes the circuit a circuit. The context of the work makes the work itself "make sense." This is the process by which a series of followed links from a Google search, for instance, can be said to tell a story; the initial search string is the initial task, the resulting "hits" are the pseudo-random subject positions, and each click of the user as he or she explores the links gradually encourages the user to see a coherence among the links—even when they occur on seemingly unrelated sites. Such circuitry acts like a carousel that one cannot step off. The effect of living as a circuitous subject, even momentarily, can be highly exhilarating because it offers self-confirmation through self-absorption. At the same time, self-confirming subjectivity makes for circuitous subjects whose inner worlds become increasingly private if left uncritiqued.

Nondialogical EEEs privatize intellectuality, and circuitous subjects created within them tend to hallucinate culture. It is difficult to resist the illusion of culture that such electronic educational environments create because they are forms of discursivity whose virtual worlds routinely model real worlds. Accustomed to regarding digital models as accurate representations of the world of our actual perceptions, comfortable users of today's virtual worlds (for instance, computer gamers) nonetheless experience "real" effects that are not public—that is, not customarily shared with others.[7] With a little ingenuity, for example, users could construct an entire history of cinema from the IMDb "My Movies" tab by creating a chronological list of films, directors, actors, and other agents and events. This might seem to a user of that database a "real" history (or at least chronology of the cinema). However, if it were published, that is, shared, it would likely be laughable because of omissions that would astonish film historians. In other words, users can (and our students often do) construct what seems to them to be versions of reality from their databases that, from a sociohistorical perspective, are so idiosyncratic that they might well be described as private fantasies.

Our perception of the boundary between the virtual and the real (between the private and the public) is affected by the technology we use. The virtuality of a tape-delayed event does not change the recorded event itself, and watching such a prerecorded event effects viewers in ways that are indistinguishable from those experienced by viewers watching the event in real time, especially if they lose the sense of the delay or are unaware of it. On the other hand, electronic educational environments that allow users to explore prerecorded culture(s) are themselves constitutive of culture; that is, it is difficult to perceive EEEs as having noticeable boundaries between virtual, private, and idiosyncratic educational experiences and actual, public, and shared ones. Both a VCR tape and a printed history book may be records of actual events, that is, of events we believe could have been witnessed by many persons; such events, then, seem public. When we watch a film like *Gods and Generals*, which includes scenes of Civil War battles, we believe these battles—though realistic—to be fictitious because of our generic expectations of films. Our generic expectations of informational media, however—broadcast news, for example—can lead us to accept fictitious images as actual ones.[8] But since most people have not yet developed generic expectations of EEEs, they have no consistent signals to help them sense the boundary between public and private readings. As a consequence, it is easy to take a virtual history (one concocted from a set of idiosyncratically motivated choices) as a real history. This is, of course, also true of print environments—the subjective selectivity of readers often results in highly idiosyncratic readings. The difference, however, is that an idiosyncratic reading is not necessarily "circuitous." A printed text is relatively linear by comparison with an EEE. Reading a novel and then going to the library and choosing another novel differs from using TIVO, for example, whose "wishlists" are predicated upon users' choices of films. Underlying TIVO is a predetermined set of subjectivities that are mapped onto each user's "history" of film watching and that are logged into a database from which the user's profile is created.[9]

We are not arguing that "reading" in electronic educational environments is fundamentally different from reading in a print environment. On the contrary, we are suggesting that they are equally subjective, but that in the EEE the nature of this subjectivity is far less well understood.[10] The circuitry of the decision-making processes in a given database positions its readers pedagogically; when EEEs have little dialogical dimension (no ongoing conversation), their circuitry fosters privatization unless a counter-initiative is simultaneously undertaken to combat it. Unfortunately, judging from the proliferation of educational CD-ROMs and DVDs, non-dialogical EEEs that reduce pedagogy to programming seem well on their way to broadly defining what electronic educational environments are about.

In our view, non-dialogical EEEs ultimately have deleterious effects that go beyond the complaints that conservative thinkers who defend canonicity make. Under the right interpellative conditions, any text's potential for "subversion" and for catalyzing cultural change can be forestalled, even when such texts are interactive and electronic. Given a sufficiently large database for example, misogynist or racist users will find it that much easier to construct "histories" that merely confirm their bigoted ideologies when exploring a non-dialogical EEE. Expansive circuitry, in other words, can actually promote circularity and insularity, a pair of characteristics that actively mask an EEE's ideological moorings.

IDENTIFICATORY SUBJUGATION

Subjugation is a danger of electronic environments, whether or not they are intended to be educational. Because of their self-confirming character, the heart of non-dialogical EEEs is "identificatory subjugation," a phenomenon that is also at the heart of the privatization of culture. From a psychoanalytic perspective, "identification" is "the process through which the subject assimilates aspects of others (objects) and constitutes its personality from the resulting products" (Harre and Lamb 86). In a section of *Rhetoric of Motives* entitled "Ingenious and Cunning Identifications," Kenneth Burke singles out "the persuasiveness of false or inadequate terms which may not be directly imposed upon us from without . . . [as in the case of interpellations], but which we impose upon ourselves, in varying degrees of deliberateness and unawareness, through motives indeterminately self-protective and/or suicidal" (559). In considering Burke's view of cunning identifications, we might also recall that

> An identification is not effected with the totality of another person or object but with some specific aspect of the person's behavior in a very specific context. The aspects of the individual's behavior that are selected for purposes of identification are congruent with or correspond to certain specific drive needs of the individual. These may relate primarily to id fulfillment of wishes, to ego's purposes of defenses, or to superego efforts directed toward self-punishment. (Arlow 137)

In other words, we usually identify not with persons but with their functions and characteristics—for example, "programmer," "critic," or "scholar" (functions) and "generous," "methodical," "overbearing," "curious" (characteristics). In an electronic environment, users tend to identify with the functions and characteristics built into the software as part of a process best likened to a circuit—a circuit of subject positions. To understand this process of identificatory subjugation, let us return for a moment to the once-futuristic application with which we began.

NewsPeek is a program that collects and organizes data from various multimedia sources. The data available in the various databanks from which *NewsPeek* gathers the "micro-texts" presented to the viewer can be thought of as analogous to the various signs that are elements of any cultural text. The user of *NewsPeek* constitutes the "macro-text" of her personalized newspaper by assembling these "micro-texts." To put it another way, each NewsPeek user constructs his or her own customized multimedia newspaper by assembling available media blocks according to the user's specified and tracked preferences. Similarly, users of CultPeeks, in constituting their personal anthologies of cultural texts, constitute their own histories of the cultures they are studying. Granted, they must use the CultPeek program itself and thus are subjected to the cultural formations implicit in the CultPeek program, but nonetheless, within those parameters, they create their own collections, which are simultaneously historicocultural meanings. The process parallels the way a reader constitutes any cultural text but is much greater in its definitive (that is, meaning-making) scope. Whereas a novel's reader manufactures that text's meaning, CultPeek users manufacture a meaning for an entire culture or cultural milieu.

Take, for example, the ways in which a student, let's call him Walter again, might be self-absorbed in reading a new text with a non-dialogical program in the CultPeek mode. Let's say he reads Alice Walker's *The Color Purple* for the first time. Because he saw Steven Spielberg's film, Walter sentimentalizes the novel. Moreover, he cannot get the image of a slimmer Oprah Winfrey out of his imagination and it interferes with his constitution of the character of Sofia.[11]

In the traditional classroom, the role private associations play in the way Walter perceives the text in relation to its historical contexts has to be negotiated in terms of its relationship to the ways in which other student readers have constituted the novel in their readings. In a book-centered classroom, interpretive strategies are evolved by interpretive communities. These communities depend for their existence upon the development of a particular history of literature from which they make sense of the text under consideration. The conventions by which a given text is made meaningful come from shared cultural histories. The same is true of dialogical EEEs. In the Lit/CultPeek and other non-dialogical EEEs to which we have called attention, however, users privately and singly constitute the contexts within which texts make sense.

In such situations, subjects are formed through self-absorption. Many people who come to a computer with trepidation soon discover that, once they learn the software, computing can be enjoyed. Even relative newcomers to computing often describe the experience of playing a game or surfing the net and suddenly realizing that they have lost track of time. In short, playing in electronic educational environments is absorbing work.[12] This, ironically, can be a problem: working in non-dialogical EEEs encourages users to "absorb" hybrid subjectivities, partly imposed by the interface (and the database behind it), and partly from the user's developing circuitous subjectivity.

This identificatory subjugation—really just part of the process of becoming a circuitous subject—can be both highly compelling and isolating. It is the condition of users identifying with their own creations of "the other" as a projection of their own interests and desires. When Walter creates a literary "history" around the figure of James Agee by investigating persons, places, and topics that have captured his interest as he browsed the database, his view of southern culture— woven from the fabric of his own desires for literature and shaped by its potential for his self-expression—is the view the multimedia database allows him. In this process, persons position themselves in narcissistic identifications to which they then are subjugated.

All EEEs—dialogical or not—teach users to behave in certain ways, of course. Non-dialogical EEEs, however, have no built-in community, a group of people who will provide "reality checks" for each other. This inherent problem with non-dialogical EEEs is not insurmountable. As with many technologies that have mixed sociocultural effects, circuitous subjectivity and identificatory subjugation can be considerably ameliorated through "subjectivity consciousness" and critique. Users (and people who work with users—for instance, teachers) can intervene in the accidental education of circuitous subjects by interrogating the circuits they've formed and the logics that they've constructed. This critique can effectively begin with the realization that EEEs always catalyze the kinds of interpellative processes we've discussed so far. Critique, in other words, is the only remedy to the self-absorption that non-dialogical electronic educational environments encourage.

Before we consider how CultPeeks are governed not only by space but also by time, we want to make one final observation about the social implications of computer programs that foster circuitous subjectivities. In our customary university habitats, when one shifts from teacher to researcher to typist, one usually walks from the classroom to the library to the office or study. In each locale, this university subject behaves differently. In an electronic environment, however, there is no sense of locale changing when subject positions change. In an electronic environment, the differences among subject positions are virtually invisible and are therefore experienced as the same. What Richard Lanham calls the "unsubstantiality" of the "electronic word" results in blurred subject positions. When persons first enter a computer world, they are not familiar with the options available to them and thus are likely to accept the role assigned to them in the program as normal. They do not notice when they become secretaries—or ambassadors—to themselves. If a professor switched positions with a secretary in the customary work environment, it would be a widely noted event. At the keyboard, however, the same change in subject positions goes unnoticed.

The process of self-absorption and privatization that we have tried to describe in this section does not stop at the boundary of the spacialized subject. Subject positions in non-dialogical electronic educational environments appear in continuous *time maps*, our term for the momentous ways in which users of CultPeek-type databases create particular histories as powerful as, if not yet as widespread, the mass media's creation of Middle Eastern cultures or Korean geography or the 1960s.

TIME MAPS

One problem with an electronic cultural history developed from non-dialogical electronic educational environments (for example, CultPeek-type databases) is the potential such technologies have for the phantasization of the world and its history. By *phantasization*, we mean the extent to which the world could become a private phantasy of the database user. Imagine an image of a world developed from television programs. Images from *Leave it to Beaver* map "the family." Images of *Cosby* map African-Americans. Images of terrorists map Iraqis while images of Rambo-esque white men map patriotism. These images could easily form a coherent mind-map of the world, and most teachers have likely encountered students who manifest such mass-media produced mind-maps: geography assembled from images on CNN; the past from TV network docudramas; the future from sci-fi movies and *Wired* magazine. All of these elements selectively and idiosyncratically combine to map out a personal worldview that encompasses both space and time.

To recap: we have been speaking of electronic anthologies, histories, and narratological experiences of particular space/times in what we have termed the "Peek" mode. Imagine now what a more extended history, a WorldPeek, might look like. In imagining that extended history, be guided by your sense of what the sixties, as an historical period in the U.S., looks like in the nostalgic and fretful return to that period in images that have populated U.S. television screens. Think of films like *The Big Chill*, newsreels of the Kennedy family, images of hippies on crime shows of the period. At present, historical data includes official records, personal memorabilia, literature and art from the period, and so on. A non-dialogical electronic educational environment's databank would be similarly constructed. In WorldPeeks, these databanks will be searched by students to form multimedia docudramas; indeed, programs like *Macromedia Flash* and *Director*, combined with image and MP3 search engines, are already making such presentations possible. Such documents will be construed, at least by some, as evidence of historical events on a parallel with the ways some network news programs docudramatize events they have not actually filmed.[13] A student using a WorldPeek databank will not differ much from Stephen Dedaelus in his spiral view of the world: me: Dublin: Ireland: the world. Spheres within spheres. A student or a teacher searching the history (possibly now quite literally a "His Story") database will likely pursue connections that are "of interest" to him. WorldPeek users would be called forth as circuitous subjects by following personal linkages through historical databanks. Moreover, they will be the very circuitous subjects to which they subject themselves. They will, in effect, invent themselves—an electronic form of narcissism—out of data that has been put in place by someone else.

Returning to the historical dimension of Peeks, consider that they will not only be world maps, that is, sets of images mapped onto a wished-for world, but also time maps. Like geographic maps, time maps show relations of proximity; that is, they suggest the temporal distance from one moment to another: U.S. slavery is an old crime against humanity, the U.S. war on Afghanistan a recent one. Also like geographic maps, time maps show relations of boundary; that is, they suggest temporal edges that mark one time as distant from another: the Harlem Renaissance is "different" from the Black Power Era.[14] In combination, these boundaried time maps begin to look like neighborhoods.

Physicists have taught us, of course, that space and time are connected and relative: U.S. slavery seems a recent crime when one knows that slavery was outlawed in England a century before the U.S. declared its independence from that country. The Harlem Renaissance and the Black Power Era seem more similar than different when both are understood as moments when African-Americans (primarily) were expending massive energies to overcome their exploitation and oppression. Spheres within shifting spheres.

The problem is that in a database, shifting spheres must become static, becoming instead categories subgrouped under categories. The dynamism of narrative history is de-animated in order to be categorized into fields, records, and tables. Users of non-dialogical EEEs built upon such databases naturally breathe new life into these limbs of history, but the revivified product is cheap. Emerging electronic cultural histories (for example, Susan E. Gallagher's *Don't Look Now*, a history of the idea of privacy in the U.S.) are made up of documentaries, not merely documents. Like most documentaries, they include varying points of view on an event that give way to circuits of identificatory positions. Extrapolating from Gerald Graff's analysis of print-based educational environments predicts that conflict, one of the motors of narrative, will dissolve in CultPeeks into undecidabilities because conflicting assumptions will, at least in EEEs arranged to appear like "news" rather than "argument," be given approximately "equal status." The consequence of this mediated treatment of conflict is that the represented conflict itself will be submerged beneath the collaborated ideology of the user and the databases to which he or she has access. This is circuitous. The CultPeek historian might say: I understand all of these positions and they are all equal in merit, identifying with all of the contradictory positions simultaneously and claiming to be at peace with the dialectic of history.

A history of rhetoric as it exists in book culture, for example, is limited by a variety of publication restrictions. Not so, an electronic equivalent. A HistRhetPeek could easily draw from a database containing all known primary and secondary sources on rhetoric—a relatively small database as such archives go. The HistRhetPeek user would simply plot a route—perhaps thoughtfully, perhaps spontaneously—through the database that fits his or her vision of rhetorical history. In other words, histories developed from Peeks would be "routes"—that is, maps of time spent in the database automatically registered by the program and readily printed as a list. To paraphrase a Microsoft interface gimmick, such a route would be a "My HistRhetPeek."

Time maps like this, since they are plotted as individual histories of decisions, tend to make history writing into records of personal phantasy. Print histories, too, are phantasized, but electronic environments not only speed up the phantasy process, they also give it a graphic, realistic, and immersive life that is semiautonomous from the daydreamer.[15] Thus a circuitous subject might not readily identify her or his integral self as being coherently—but illusorily—located in a self-constructed time map of a world because components of that world can be documented in a variety of more traditional (and more generally accepted) media: movies, timelines, and hard copy. Also troubling is that this daydreamer-turned-historian might not realize that subtle shifts had taken place in her or his role in the phantasm-world or that the functions of the subject positions in her or his customary world had also subtlety changed. Our Walter, for example, might not notice how much more labored—in the Marxist sense—his work had become, how much more intellectually powerless and less creative his new work was turning into. He might not notice how his demeanor had changed. Unaccustomed to thinking of himself as an authoritarian, he might be surprised to find how restrictive his computer-assisted history turned out to be. In short, Walter might wake up one day a completely different subject—and never know it.

CONCLUSION

We can probably halt the privatization of culture already underway by taking an active interest in how electronic educational environments are programmed. For example, EEEs can be designed as features of Local Area Networks and integrated with dialogical software to maintain the social (dialogical) dimension of cultural study. Here, then, lies the main difficulty—for now, at least: Who will program our future electronic educational environments? Will our hypertext books always be prerecorded? If we give over the programming of EEEs to educational businesses and their university cohorts, might we be condemning ourselves to becoming circuitous subjects whose self-absorption will have an alienating end?

These reflections underscore the motives for designing projects like Alternative Educational Environment's ASCEND project (www.ascend.comm.uic.edu); the Virtual Harlem, Virtual Montmartre, and Virtual Bronzeville Projects (www.caee.net); and the *Thirst* Project, which is aiming to use a commercial computer game engine to implement digital game-based learning in college classrooms (lgi.mesmernet.org). Marsha Kinder's "Labyrinth Project" (www.annenberg.edu/labyrinth) is also a useful example of what can be achieved through innovative database narratology, as are Lucy Petrovich's cooperative digital art installations (www.arts.arizona.edu/lucy/bio). Such projects offer useful guideposts to anyone committed to stemming the tide of non-dialogical electronic educational environments, and interested in developing more dialogical alternatives. In our experience, the best high-tech alternatives to circuitous subjects in their time maps begin with no technology at all. Rather, they begin in community, in developing subjectivity consciousness, and in the sensitivities enabled by the recrudescence of one's identity as they mature in electronic environments.

NOTES

1. The Committee on Virtual Reality Research and Development was convened in 1992 by the National Research Council "at the request of a consortium of federal government agencies: the Advanced Research Projects Agency, the Air Force Office of Scientific Research, the Human Research and Engineering Directorate of the Army Research Laboratory, the Crew Systems Directorate and the Human Resources Directorate of the Armstrong Laboratory, the Army Natick RD&E Center, the National Aeronautics and Space Administration, the National Science Foundation, the National Security Agency, and the Sandia National Laboratory" (Durlach and Mavor vii).

2. While one might take a different tack than the one we offer in this essay and argue that if it's okay for actors to perform roles based on scripts they have not written then it should be okay for users of EEEs to do the same, as this essay

will elucidate, our response to such an argument would be that actors are conscious of the fact that they are performing roles invented by someone else, while users of EEEs often are not.

3. Such access is already available in certain parts of the U.S., as well as in many parts of Scandinavia and Japan.

4. The virtual dialectic, modeled as it is on the conceived parameters of the real-word, is therefore necessarily bound up with the real-world dialectic.

5. There are many comparable—and more realistic—roles that computer users inhabit: sys admin, moderator, and CyberAngel, are just a few.

6. Such functionaries can include particularly enthusiastic users who have added on to the virtual world in their free time.

7. Such virtual worlds probably should be regarded neither as fictional nor as real, but rather as existing on a continuum stretching between the end points of "common" and "uncommon" experience.

8. The most striking recent example is Jayson Blair, the *New York Times* journalist who was recently discovered to have invented many of the details of his stories.

9. This necessarily and simultaneously constitutes a form of surveillance. See Pogue; Harmon.

10. As a cognitive process, reading a printed text and reading in an EEE is not "fundamentally" different. However, as a "learned practice" or habit of reading, they differ considerably. See Sosnoski.

11. In *Teletheory*, Greg Ulmer proposes this very scenario, terming it a "Mystory" (rather than a "history"), as a curriculum for the future. See 44 ff.

12. A more detailed discussion of how electronic environments—particularly computer games—require different types of work can be found in McAllister, Ruggill, and Menchaca.

13. Perhaps the most common recent examples of such docudramatizations were the plethora of computer-generated animations of various aspects of the U.S. led wars in Afghanistan and Iraq. With the government's censorship of front line coverage, many U.S. news organizations were forced to stitch together media clips provided to them by the Department of Defense (the result of which can reasonably be called "propaganda"), or if their budgets permitted, they generated 3D computer models of their stories' substance: ground force deployments, ordnance effects, weather and terrain problems, and so on. In both cases, such docudramatizations are fictions based on (purported) facts and passed off as very real looking "news."

14. Recent scholarship has begun to problematize the notion of "periodization," linking it to imperialism, colonialism, and hegemony. See, for example, Said, Harvey, and Spivak.

15. It may be argued that the privatization we describe vis-à-vis non-dialogical EEEs also takes place in book-oriented learning environments and that they are not different in kind but only in degree. We have two answers to this objection: first, while we agree that privatization occurs in both instances, we contend that the "speed" (degree) of the privatization differs; second, in some matters, speed changes the relationship between kind and degree. Cream, when

whipped, has a very different effect on one's appetite for it. This is not a small matter. Our appetite for privatization is already keen. Finding privatization to be a good (the object of an appetite) would be a substantive change for many, and could arguably be termed a new form of "isolationism." Such a tendency is counter to the cross-cultural exchanges we appreciate and work to encourage.

WORKS CITED

Althusser, Louis. *Lenin and Philosophy and Other Essays*. Trans. Ben Brewster. New York: Monthly, 1971.

Arlow, Jacob A. "Object Concept and Object Choice." *Essential Papers on Object Relations*. Ed. Peter Buckley. New York: New York UP, 1986. 127-46.

Brand, Stewart. *The Media Lab: Inventing the Future at MIT*. New York: MIT P, 1987.

Burke, Kenneth. *A Grammar of Motives and A Rhetoric of Motives*. Cleveland: Meridan, 1962.

Durlach, Nathaniel I., and Anne S. Mavor, eds. *Virtual Reality: Scientific and Technological Challenges*. Sponsored by the National Research Council's Committee on Virtual Reality Research and Development. Washington: National Academy of Sciences P, 1994.

Eco, Umberto. *A Theory of Semiotics*. Bloomington: Indiana UP, 1976.

Gallagher, Susan. *Don't Look Now: A Documentary History of the Idea of Privacy in the U.S.* DVD. http://faculty.uml.edu/sgallagher/privacydescription.htm. 2002. (9/10/2005).

Graff, Gerald. Professing Literature: An Institutional History. Chicago: U of Chicago P, 1987.

Harmon, Amy. "TiVo Plans to Sell Information On Customers' Viewing Habits." *New York Times* 2 June, 2003.

Harré, Rom and Roger Lamb. *The Dictionary of Developmental and Educational Psychology*. Cambridge: MIT P, 1986.

Harvey, David. *The Condition of Postmodernity: An Enquiry Into the Origins of Cultural Change*. Cambridge, MA: Blackwell, 1990.

Kenny, Anthony. *Action, Emotion, and Will.* London: Routledge, 1963.

Lanham, Richard A. "The Electronic Word: Literary Study and the Digital Revolution." *New Literary History* 20.2 (1989): 265–90.

McAllister, Ken S., Judd Ruggill, and David Menchaca. "The Game Work." *Communication and Critical/Cultural Studies* 1.4 (2004): 297–312.

Pogue, David. "State of the Art; For TiVo And Replay, New Reach." *New York Times* 29 May, 2003.

Ruggill, Judd Ethan. "Corporate Cunning and Calculating Congressmen: A Political Economy of the Game Film." *TEXT Technology* 13.1 (2004): 53–72.

Said, Edward. *Covering Islam: How the Media and the Experts Determine How We See the Rest of the World.* New York: Pantheon, 1981.

Sosnoski, James J. "Hyper-Readers and Their Reading Engines." *Passions, Pedagogies, and 21st Century Technologies.* Ed. Gail Hawisher and Cynthia Selfe. Logan: Utah State UP, 1994. 165–71.

Spivak, Gayatri. *A Critique of Post-Colonial Reason: Toward a History of the Vanishing Present.* Cambridge: Harvard UP, 1999.

Ulmer, Gregory. *Teletheory: Grammatology in the Age of Video.* New York: Routledge, 1989.

7
TOWARD A RHETORIC OF NETWORK (MEDIA) CULTURE

NOTES ON POLARITIES AND POTENTIALITY

Byron Hawk

> The desire for simplicity has long inspired efforts to explain the world in terms of simple systems that function smoothly and simple laws that can be reduced to simple equations.
> —Mark C. Taylor

The desire for simplicity has haunted rhetoric and composition for most of its history, from stock forms for producing oral speeches in ancient times to simple processes for the production of written texts in contemporary times. Even though the desire to utilize simplicity to understand the world was the driving force of science, Mark Taylor recognizes that science is at root a metaphysical if not theological enterprise (137). Simplicity is always abstracted from complex realities. This has certainly been the tendency in rhetoric and composition, whose primary debate has been between two opposing methods for simplifying the complexity of writing. On the one hand, expressivists responded to writing's complexity by abandoning system altogether. On the other, rhetoricians of various stripes have tried to produce simple systems that make writing teachable. Neither faction has fully addressed the complexity of writing, and I think the value of Taylor's book is that it may lead the field to confront this reality. Though some of his conclusions regarding education in general are debatable, the book does a good job of summarizing much of the work in complexity theory and applying it to various fields—art, architecture, theory, and communication. One of the fields that Taylor does not directly address, however, is rhetoric. Just as rhetoric and composition is currently confronted with the complexity of writing, rhetorical studies is in the process of trying to determine just what rhetoric would be in our current cultural situation. The ancient civic space that led to the emergence of rhetoric has been replaced by contemporary network space. In its place, however, are few rhetorical theories that adequately address the complexities of this new social space. Simply applying rhetorical systems developed in the context of ancient Greece to our contemporary period seems to fall into the desire for simplicity that Taylor hopes to counter.

Consequently, rather than produce a counter-response to Taylor, this essay aims to extend his work into rhetorical studies by examining certain traditional rhetorical concepts in relation to concepts articulated in Taylor's book. In order to lay some initial groundings for a rhetorical theory based on the topoi of complexity and networks, it is important to create compositions or polarities between key terms: Heuristics : Schemata; Dissoi-Logoi : Polarities; Rhetorical Situation : Complex Adaptive Systems; Kairos : Emergence; Logos : Network; Ethos : Screen; Pathos : Affect; Process : Evolution. These initial linkages cannot provide a fully fleshed out rhetorical theory in a short response, but instead enact strange loops, set ideas in motion. My purpose is to suggest the potential relevance of Taylor's work to contemporary rhetorical theories and to call for further inventions in the areas of new media environments and network cultures.[1]

HEURISTICS: SCHEMATA (GRIDS)

A heuristic is a set of questions, a mental grid, or a generic process that aids its user in inventing and articulating ideas. The writer sets this grid between him or herself and the world in order to impose order on its chaos. Though most who discuss heuristics tout their openendedness (each enaction will generally produce unpredictable results), in practice they inevitably function as grids—the heuristics themselves remain unchanged. Taylor's use of "schemata" revises this grid-lock. The grid maintains both stability and simplicity—the heuristic questions remain the same or the tagmemic grid remains the same (23). But schemata actually move such mental grids from a synchronic position into an evolving process. Schemata change in response to input: "Emerging schemata identify, compress, and store the regularities of experience in a way that makes it possible for the system to adapt by responding quickly and effectively" (206). In other words, we start with experience, generalize a pattern or schema from that experience, turn that pattern on future experience, and then adapt the pattern to devise a new schema. Taylor turns this position/process onto the modernist notion of grids, revising it for a postmodern context. Using Chuck Close's painting as an example, he shows how a grid can actually produce work with greater complexity. Each cell in the grids comprising Close's paintings contains an abstract painting in itself.

Each individual cell-painting combines to create the effect of a larger work, thereby emulating a network logic where the whole extends beyond the sum of the parts. As the example of grids shows, a schema (like the notion of a grid itself) is caught up in complex networks that evolve and adapt to new circumstances. This basic process has implications for rhetorical heuristics: (1) students need to develop their own schemata to fit their particular topics/situations, and (2) if we give them schemata first, their goal should be to revise those schemata as a part of the invention process rather than follow them prescriptively.

In "The Meaning of Heuristic in Aristotle's Rhetoric and Its Implications for Contemporary Rhetorical Theory," Richard Enos and Janice Lauer attempt to interpret heuresis as more than techniques used "to find out or discover." They argue that the term also means to create meaning through language, not just to discover what already exists. Heuristics under this reading enable the rhetor to create probable judgments—to assess a situation and co-create meaning with the audience. Though Enos and Lauer hint at the possibility of producing "entirely new proofs" beyond "existing topoi," they are firmly situated in the autonomous subject who makes such a rhetorical choice (82). Their level of complexity doesn't really move beyond a rhetor/audience dialectic for producing socially constructed meaning. They step back from a purely epistemic approach by noting that Aristotle "considers empirical investigation and syllogistic reasoning as processes of thinking that are not necessarily discursive" (83). But even this nod to reality doesn't move them beyond the simplicity of the communications triangle, the linear movement of solving social problems, and the turning back to culture (and language) as the primary medium. Gregory Ulmer's *Heuretics: The Logic of Invention* probably comes closer to addressing the implications of Taylor's work. Ulmer's use of heuristics sets out to make "students become producers as well as consumers of theory" (xiii). His heuristic becomes a method for inventing new methods through the loss of the subject in a complex system of discourse and the world. His grid (whether the CATTt or the popcycle) folds over into the new by immersing his students in electronic culture and asking them to map their own trajectories through the territory. Each specific node that they encounter or inhabit is a new cell that once collected together in a set, or a new grid, functions to produce a new whole, a new conception of the complex situation. Enos and Lauer are examining and using Aristotle's heuristics/topoi; Ulmer invents his own and asks his students to do the same.

DISSOI-LOGOI: POLARITIES (STRANGE LOOPS)

Dissoi-Logoi is an ancient rhetorical exercise based on making the weaker argument the stronger, or reversing the obvious argument to make an argument that is culturally/situationallycounterintuitive. The exercise was meant to unfreeze rigid, accepted concepts/positions in the same way Taylor turns to the notion of the grid. Taylor's recurring position is that such oppositions are not really competing oppositions—each side exists because of its relationship with its "opposition." Rather than competing oppositions that produce a winner or synthesis, they are polarities, or nodes caught up in co-producing systems. Baudrillard, according to Taylor, argues that the oppositions implode—one pole, such as simulacra, engulfs its opposite, the real. But for Taylor, these polarities are caught up in "strange loops": "Strange loops are self-reflexive circuits, which, though appearing to be circular, remain paradoxically open" (75). Oppositions, then, are polarities caught in these complex relationships in which each side is evolving and changing in relation to the other. Polarities, like cells in a grid, don't function alone but in complex, co-adaptive relations with other polarities, creating a larger whole. One pole isn't overtaken by the other. It changes and thereby changes its opposition. Any rhetorical system that uses opposition or dialectics to privilege one pole or the other, such as James Berlin's heuristic, should recognize the ecological, co-productive nature of the "weaker" argument that can easily turn to become the stronger. Situations are more complex than dialectics accounts for.

In *Rhetorics, Poetics, Cultures*, Berlin constructs a heuristic that identifies key terms, sets up an opposite term, and then prompts the students to value one term over the other (126–28).[2] Such a strategy follows the model of *dissoi-logoi* articulated by John Poulakos. Poulakos starts with the basic Protagorean position that on every issue there are (at least) two opposing arguments. Such a worldview is founded on difference rather than unity and is consequently fundamentally rhetorical. At any point on any issue a contrary argument can be found and put forth. But in order to act, people must be persuaded to one side or the other, even if temporarily. Poulakos gives an example from Prodicus: in the story of Heracles at the Crossroads both Vice and Virtue argue for their means to happiness. Both make arguments for their positions and against the other's, but in the end Heracles still makes no decision. Poulakos reads this as leaving the choice up to the reader, who in similar circumstances will have to choose: "since the imperative to action demands that an impasse be overcome and that a choice be made, the human subject must in some way disturb the balance of perfectly opposed alternatives. This means that in the final analysis one must prefer one option over all others" (59; emphasis added). Such a position on *dissoi logoi* sets aside the possibility that action can occur even without such a choice based on the traditional model of the subject. If an alternate position is always on the horizon, why not work toward another, and yet another? This position is put forth by Debra Hawhee in "Kairotic Encounters." She notes that in Gorgias' Helen his approach to *dissoi logoi* doesn't simply put forth two possible arguments, but several possible reasons Helen might not be responsible for her own actions: "Gorgias does not settle on one definitive explanation, but enumerates several viable ones" (26). Helen acted not because she chose between two options but because multiple forces moved her. Following Deleuze and Guattari, Hawhee notes, "The movement of Gorgias' speech, then, occurs in the middle, in the realm of the between. [. . .] Gorgias' betweenness, not necessarily Gorgias himself, seeks not to replace the previously accepted 'truth' about Helen with another truth, but rather to undermine the very notion that one truth [. . .] exists" (27). This logic of the "and" of an always expanding *dissoi logoi* situates the subject in a network of multiple forces and attempts to address the complexity of our current cultural situation.

RHETORICAL SITUATION: COMPLEX ADAPTIVE SYSTEMS

The notion of the rhetorical situation has been an important concept in modern rhetoric because it turns toward an emphasis on social construction, or the recognition that situations generate discourse. Its early development as a concept participated in a generative polarity: Lloyd Bitzer argued that the situation causes discourse as a direct response and Richard Vatz countered that discourse is always already a part of the situation and can thus create or determine situations. The polarity of context-text operates as a strange loop, and the combination of multiple strange loops Taylor calls a "complex adaptive system." Both biological and cultural complex adaptive systems remain open to their environments and adapt accordingly. Never static, they produce larger scale behavior, texts, and structures from the movement and interactions of smaller parts. Individual ants, a recurring example for Taylor, form the larger entity of the colony, even though the ants are unaware of this larger whole. The ant colony can react to environmental conditions and adapt/evolve accordingly, even though the colony has no equivalent of a "mind" at the level of the whole. Each ant reacts only to its immediate neighboring ants and circumstances but the larger flow of the colony nevertheless has a coherent, complex movement. Such complex adaptive systems produce strange loops among their individual parts that create "effects disproportionate to causes"—the interaction of the individual parts will at some point reach a "tipping point" or a "moment of complexity" where their interaction and feedback loops produce a qualitative change at the level of the whole (165). In order to do this, these self-organizing systems have to process information in a way that goes beyond simple reaction. They also adapt. Schemata become critical elements in this development. A complex adaptive system has to (1) identify regularities in its environment; (2) generate schemata (or models, theories) that enable it to recognize these regularities; (3) have schemata that adapt to the changing circumstances as well as in relation to other schemata (different schemata in a complex adaptive system form a sub-network or complex adaptive system within a complex adaptive system); (4) have schemata that can predict environmental activity (as with evolution, reliable schemata continue to function, unreliable ones change or disappear); and (5) tie into its environment through feedback or strange loops (166–68).

The strange loop created by Bitzer and Vatz is but one polarity in such a larger system. Barbara Biesecker attempts to get beyond this initial polarity to a larger complexity in her contribution to the rhetorical situation debates. She questions Bitzer and Vatz's assumptions about the autonomous elements of the rhetorical situation and the causal logic that establishes the relationships among them through Derrida's notion of differance. Bitzer presupposes a causal chain from reality to speaker to text to audience to reality, positing the initial exigence as the origin of the causal chain. Vatz, she argues, simply starts the chain with the speaker who through interpretation of an exigence puts his intention into the situation via discourse. The "Bay of Pigs," for example, was created as an event because of the decision to create a particular type of discourse about it, which turned an "event" into a "crisis." In other words, the causal relationship is reversed: rhetoric becomes a cause not simply an effect. The problem for Biesecker, of course, is the simplicity of this model. She initially problematizes the notion that an origin stems from an autonomous event or speaker. For her the originary moment is not in the situation or speaker but in their difference, in the absent space between them created by their relation. Rather than an event or speaker initiating a relation, their emergent relation co-produces each one through an ongoing development. Her example is Derrida's *Glas* which enacts differance by placing two columns of text side-by-side in the same book. The meaning of the text is not in the left side on Hegel or the right side on Genet. It is in the absent space between them, in all the possible relations and connections between them that can arise during a reading of the text. Such a perspective displaces subject and event and "refuses to think of 'influence' or 'interrelationship' as simple historical phenomena" (Spivak, qtd. in Biesekcer 121). If a speaker's subjectivity is formed through such a moment of differance, the same can be said of audience, not to mention language and the world.[3] All the elements of a rhetorical situation are effects of their place in an economy of differences—they each form polarities with the others and evolve co-adaptively. Biesecker is after emergence and is using the Derridian lexicon to find an articulation of it. She comes close to the notion of emergence but doesn't move all the way to something like complex adaptive systems. Environment, rhetoric, texts, and audiences are complex adaptive systems in themselves and together form other complex adaptive systems. What we have are networks linked to other networks. Complexity theory can give us a new set of terms beyond the notion of differance through which we can understand and articulate the complexity of rhetorical situations, especially in a complex media-rich environment where so many events are co-produced through media.

KAIROS: EMERGENCE

Kairos is the classical rhetorical term for chance and timing—both the situation's apparently random ability to seize a rhetor and the rhetor's ability to recognize the right discourse for a given situation. Rather than pure chance or chaos, kairos is complex and requires the rhetor's ability to participate in the co-adaptive development of a situation by infusing discourse into it. Taylor regularly makes the distinction between chaos and complexity. Chaos theory was developed in opposition to Newton's mechanical physics and examines situations that cannot be conceptualized (schematized) due to the lack of information. Complexity theory examines situations that emerge between the polarity of chance and order where "self-organizing systems emerge to create new patterns of coherence and structures of relation" (24). Polarity is essentially a balancing act between too much order and too little order, but it is this dynamic that fuels emergence. Emergence refers to that moment of complexity when the interaction of parts or system components generates unexpected global properties not present in any of the local parts. Microscopic and macroscopic systems operate in loops that have both negative and positive feedback. Negative feedback turns the balance toward equilibrium, which shuts down the movement of the system. Positive feedback interrupts equilibrium by increasing both speed and heterogeneity. Speed increases the interaction among parts and increased interaction creates more diverse components—more diverse components move the system from linearity and stability to recursiveness and complexity. These two properties can give rise to effects disproportionate to the immediate causes, thus producing emergence (143). The polarity of order and chaos also tends to follow a particular sequence from order to complexity to chaos to complexity to order (146). The transitions in this sequence are generally unpredictable, though we can determine them in retrospect. As they are happening it appears as if chance is a predominant cause that accounts for "effects disproportionate to their causes," but in retrospect we can determine many of the patterns of complex activity and interrelatedness of the system parts. Emergence, then, operates in the moments of complexity between the polarity of chance and law (149). It is this point of emergence that signifies kairos.

Carolyn Miller notes that there are "two different, and not fully compatible, understandings of kairos" (xii). One places the importance on the rhetor, choice, and decorum (adapting to what is already culturally established as appropriate for the situation); the other places importance on "the uniquely timely, the spontaneous, the radically particular" (xiii). This position asks the rhetor to be creative in the face of human life's "lack of order" (xiii). E.C. White takes up the latter in his seminal work *Kaironomia*, which addresses the notion of emergence. Noting that kairos traditionally meant the right moment or opportune time, White argues that we can never manage present opportunity, even provisionally. Success depends on "adaptation to an always mutating situation. [. . .] Such an activity of invention would renew itself and be transformed from moment to moment as it evolves and adapts itself to newly emergent contexts" (13). White uses the concepts of *dissoi-logoi*, or polarities, and a Heraclitean worldview of flux and becoming to theorize the emergent space between polarities. We end up with a polarity of rhetor and context. A rhetorical choice in the moment only solves the "tension between contrairies" by a force of will in the hope that chance produces an utterance that has meaning within a situation (16). Again, I think complexity theory can give us a language that articulates the two apparently incompatible conceptions of kairos—one that favors the rhetor and one the situation—within a single system. Seizing the moment means being able to anticipate it, unconsciously as well as consciously, not just reacting to it but adapting to it, with it, and often times quickly. We may never be able to completely predict complex behavior, only recognize elements of it in retrospect, but being able to recognize more of it as it is developing and to understand how we might co-adapt along with it will become a vital rhetorical skill. As Miller notes, "The most complex and interesting rhetorics, both ancient and contemporary, include both dimensions of kairos in some way, keeping them in productive tension" (xiii). It is just such a polarity that can be fruitfully theorized for contemporary media culture via complexity theory.

LOGOS: NETWORK

Logos in traditional rhetorical terms refers both to a narrow notion of logic that generally follows the enthymeme, a fairly simple system, and to a larger conception of language and the power of the word, a fairly abstract or chaotic system. For Taylor, of course, "far from opposites, simplicity and complexity [...] are braided 'like hair intricately tressed and knotted.' Such knots create binds and double binds that transform seemingly simple questions into exceedingly complex puzzles [... and] we [have] become ever more deeply enmeshed in the logic of networks" (199). Logic, Taylor argues, is not just the imposition of simplicity, linearity, and system onto the world or the chaotic power of language but also the knowledge that emerges from networks of relations, complexity, and noise (200–01). The simple sequence of emergence outlined above isn't really a sequence; it functions through a network logic, which has three basic characteristics: its basic structure is a set of nodes (or knots) and the relations among those nodes; its basic dynamics are determined by the strength of the connections or relations; it "learns" or evolves via the changing strengths of the relations that adapt to the nodes around it (154). Network logic operates through nodes that communicate with one another. These nodes or knots work like switches or routers that send, receive, and transmit information. As Taylor puts it, "The ways in which connections intersect create the distinctive traits and functions that differentiate nodes. While the connections of each node ramify throughout the network, the relations that are most decisive are relatively localized. [...] [I]f there are too few connections the network freezes, and if there are too many it becomes chaotic. Since the interrelations of nodes are both reciprocal and many-to-many, feedback loops can be both positive and negative" (154). Networks, then, are decentered and do not have to operate on ordered, "logical" sequences but run in parallel and recursive sequences. Such a logic means that networks become complex adaptive systems because there are multiple networks co-adapting to one another (171). The seemingly simple, static logic of the enthymeme and the abstract power of language over us need to give way to a more complex middle-ground. If we want to understand the way language functions in complex (media) economies, we need a logic, a new image of logos, based on the network not as a static system but as a system in motion.

Logos is a very complex term. G.B. Kerford notes that the term caries at least three levels of meaning which refer to three different applications or uses: language (speech, discourse, argument, description, statement; thought), mental processes (thinking, reasoning accounting for, explanation); and world (structural principles, formula, natural laws, which "are regarded as actually present in and exhibited in the world-process") (83). In other words, the term implies the larger complex in which bodies, texts, and thought are always connected.[4] Kerford warns that whenever we see the term in the pre-Socratics, Aristotle, or the sophists, it always carries elements of all three meanings. Thinking in terms of this larger connotation, it is easy to see why someone like Victor Vitanza would place emphasis on the power of logos. In *Negation, Subjectivity, and the History of Rhetoric*, he sees logos both as a system whose power excludes as well as a force that continually resists systematization: logos as hememony and logos as dynamis (127). These two elements form a polarity that is never static. Following Heraclitus, Vitanza equates logos with strife, the ongoing movement of logos as law, discourse, custom against but alongside logos as world, force, movement. No one side wins this battle. It is the ongoing movement that produces complexity or heterogeneity, or what Vitanza calls *dissoi-para-logoi* (111).

The energy produced by the force of the tension in a such polarity breaks out into multiplicity. Multiplicity, I would argue, operates on the logic of the network. As Taylor notes, network is not a static, frozen logic. It is in motion, it adapts, it never stands still. In *The Semiotic Challenge*, Roland Barthes explores the territory of rhetoric both diachronically and synchronically. Conducting a journey through the history of rhetoric, he collects fragments of rhetorical discourse to produce a network that folds over to produce more discourse (15–16). Movement through networks creates new links and new networks to be traveled through and re-linked. Likewise, complexity theory attempts to combine a diachronic and synchronic perspective—rhetoric becomes a system that moves and evolves. As Lanham argues, the Greek word *logos* meant something more like *information*: "Life is information; life is logos. It is an evolutionary system, dynamic, perpetually emergent. It creates new meanings, [. . .] rather than simply communicating preexisting knowledge in a transparent capsule" (254). In this system, the driving force is noise—that which the (static) system cannot account for but which forces the system to move, rearticulate, and reconnect. If logos as word is becoming less relevant in contemporary media culture, it doesn't mean logos is dying. It is simply adapting to new situations, highlighting other elements in the complex meaning of the term.

ETHOS: SCREEN (NODE)

Ethos is the classical concept of character, or the identity of a person as exhibited to an audience or social group. It is generally linked to the modern notion of the self or subject, but ethos implies no such inner morality. Rather, it signifies an ethics that is relative to the rhetorical situation. For Taylor, "In a network culture, subjects are screens and knowing is screening" (200). A screen is something that divides an individual from its environment; it protects or conceals or screens out material, while at the same time allowing certain outside elements through. It divides and links. Knowledge is gained by screening or filtering the noise of the world and developing schemata to negotiate the surrounding environment. This allows the particular body or node in question to link or connect to its local situation. In our contemporary situation, the excess of information creates an excess of noise and the increased demand for schemata or screens. A set of screens comprises a node, a point of connection in a network. As Taylor notes,

> The self—if, indeed, this term any long[er] makes sense—is a node in a complex network of relations. In emerging network culture, subjectivity is nodular. Nodes, [...] are knots formed when different strands, fibers, or threads are woven together. As with the shifting site of multiple interfaces, nodular subjectivity not only screens the sea of information in which it is immersed, but is itself a screen displaying what one is and what one is not. (231; emphasis added)

Taylor's pun on the term *screen*, from partition to sieve to computer monitor, elaborates on the social aspects of ethos. No longer linked to an inner subjectivity (and abstract morality), who we are is how we are linked and presented to the surrounding ecological situation. This "self" as node emerges from the screening process "without any centralized agency or directing agent" (205). Therefore, any account of the subject in a contemporary rhetorical theory for media culture cannot presuppose an interiority (at the very least one that preexists its situatedness), linking the concept of ethos to an ethics that is based on a relational, network logic and complex adaptive systems.

James Baumlin, in his introduction to *Ethos: New Essays in Rhetorical and Critical Theory*, notes that "the etymology of the term *ethos* invites such an opposition [between a central self and a social self]. Translated as 'character,' ethos would seem to describe a singular stable, 'central' self. Translated as 'custom' or 'habit,' ethos would describe a 'social' self, a set a verbal habits or behaviors, a playing out of customary roles" (xviii). He links these two positions to Plato and Aristotle respectively. Plato's emphasis on truth and the individual predisposes this position toward a concept of morality. Aristotle, on the other hand, is interested in "an active construction of character" (xv) based on ethics. For Aristotle, the rhetorical situation makes the speaker an element of discourse, not its origin—ethos becomes the product of delivery within a specific situation (xvi). In *The Use of Pleasure*, Michel Foucault takes up this distinction between a central self and a social self, arguing that the Greeks generally followed Aristotle's social self. For the Greeks, a man's (and, of course, it is man's) specific acts aren't at issue. If a man wants to have credibility within the community, he is expected to act with moderation, no matter what type of relation he establishes. The Greeks are not interested in acts, as the Christians are with their pastoral notion of the flesh, nor in an inherent sexuality per modernity, but in the dynamics of the act-pleasure-desire set of relations. This ensemble is fueled by the force of desire. Nature provides pleasure in an act/relation; pleasure gives rise to desire; and the desire for pleasure leads to a repetition of the act (42-43). The ethical question becomes one of the man's use of pleasure. Rather than follow an abstract morality based on a central self, men are expected to operate under self-mastery with regard to their desires and the relationships they establish based on their desires.

Though this situation sets up the further historical development of the subject, our contemporary period signals the subject's absence.[5] The image of a single, central, stable subject gives way to a multiplicity of selves. This goes beyond Aristotle's custom, or dramatis personae, to the multiplicity of relations we establish. In a media culture this distinction is vital. It is not simply the identity we might try out online but the relationships established through those identities and the affects of those relationships on bodies. Gilles Deleuze turns to Spinoza to find an ethics for this ethos. Such an ethics doesn't look to a subject but to a body and how that body sets itself into relations or compositions with other bodies. The body's ethical goal is not only "to actualize its potential to increasingly higher degrees" (Massumi, Users 82), but also to actualize and increase the potential of the other bodies within that relation.[6] Such an ethics/ethos of networked relations will be a key to future understandings of rhetoric in networked, media cultures.

PATHOS: AFFECT

Pathos is one of the key concepts in classical rhetoric that distinguishes rhetoric from logic. Human minds don't just screen information logically, they respond to information emotionally. Traditional rhetorical theories put most of the emphasis on psychological theories of emotion or character types or on ideological assumptions.[7] But as many contemporary cultural theorists have recognized, especially Gilles Deleuze and Brian Massumi, bodies also respond "emotionally" not just minds.[8] Taylor doesn't talk explicitly about affect (or Deleuze) in *The Moment of Complexity*, which is one of the book's shortcomings. Without this concept, his system makes an incomplete (rhetorical) theory. Charles Taylor, recognizing affect's importance to theories of language, notes that Herder is one of the first to see that the acquisition of language created a new reality, not language as representation of objects but language as the production of affect/consciousness. Language produces "not just anger but indignation; not just desire but love and admiration" (105). This new affective dimension to life makes possible new sets of relations—"intimacy and distance, hierarchy and equality" (106). This concept of affect is based on using language to create affective relations among singularities or nodes. The body is essentially a node with multiple screens: language produces new screens and schemata that affect the body's links with the environment and other bodies, allowing us to adapt, to form new relations, connections, and networks. Pathos is about using ideas/feelings in an audience to ground persuasion or about creating those emotions in an audience. But affect moves us toward relations among bodies, which is critical to understanding (discourse in) network culture. Like language, new media make new affections and new relations possible.

The typical misstep is to equate affect with emotions. For Massumi, "An emotion is a subjective content, the socio-linguistic fixing of the quality of an experience which is from that point onward defined as personal" (*Parables* 28). Affect, on the other hand, operates at an a-subjective, pre-linguistic level. Charles Taylor's notion of affect is still operating through language, though he recognizes the key element of opening new potential relations. But Massumi's affect is a bodily experience that "lies midway between stimulus and response": it is that point at which the body is enacting multiple relations below its cognitive ability to perceive them; the body is "bathing [in] relationality" but not consciously accounting for every molecule of water (61). Massumi recognizes the key connection to complexity theory:

> It is all a question of emergence, which is precisely the focus of the various science-derived theories that converge around the notion of self-organization. [. . .] Affect or intensity [. . .] is akin to what is called a critical point, or bifurcation point, or a singular point in chaos theory and the theory of dissipative structures. This is the turning point at which a physical system paradoxically embodies multiple and normally mutually exclusive potentials. [. . .] (32)

Affect, then, is "an ability to affect and a susceptibility to be affected": it is a body's capacity for relations within a network—the potential for linking to and being linked to. Emotion is simply "recognized affect" (61). Operating outside of a consciously recognized personal or subjective mode, affect functions at the level of node/screen. The body may screen out some potential relations while filtering in others, keeping it open to relations. A node is not unified or self-contained but multiple and fragmented, connected to the world through various affective relations. Because media operate as additional screens for a node, any understanding of rhetoric in media environments needs to understand rhetoric at the level of affect.

PROCESS: EVOLUTION

The writing process (prewriting, writing, rewriting) is credited with bringing rhetoric and especially invention back to the forefront of composition. In many ways it has performed this function well, linking the frozen product of writing to the larger context that produces it. However, process theorists have never fully considered the connections between evolutionary processes and thinking/writing. Taylor explicitly makes these connections, arguing that the writer as screen operates in a polarity with the situation and in an economy of personal experience, texts that are read, and words that are written. In this context, the writer reaches evolutionary roadblocks (schemata no longer fit the circumstances, forms are no longer relevant). These evolutionary dead ends change schemata, which changes possible relationships, and thereby affects the larger evolutionary development. The writer becomes a circuit relay in this larger economy: "Words, thoughts, ideas are never precisely [our] own; they are always borrowed rather than possessed" (196). Just as bodies are the vehicles for genes that live on after bodies pass away, writers perform a host function for ideas in our cultural ecologies. Writing, though, has another layer of complexity: "Rewriting does not merely repeat but also transforms in a way that complicates the host/parasite relationship" (196). The text is at one point in the process a parasite on other texts, but during the process it reaches a "tipping point" and transforms into a host that others will enter into a parasitic relationship with and ultimately transform. This parasitic complexity problematizes any simple relationship to time. Thinking/writing not only has "rhythms of its own" but it is also "impossible to know just how much time is required for thought to gel because [the writer is] not in control of this process" (197). Though evolutionary time appears linear, "[t]he time of writing does not follow the popular figure of the line because present, past, and future are caught in strange loops governed by nonlinear dynamics" (198). Like ants, writers are a "colony of writers" caught up in the larger evolutionary flows of other networks. In short, "[t]he moment of writing is a moment of complexity" (198). And for Taylor, we write to produce and embody these moments in order to contribute to the evolution of thought/schemata and thus the whole (cultural) ecology. Our writing should disturb, create more noise, push equilibrium into new relations and assemblages (198).

The writing process, while starting out as an attempt to bring movement and recursivity to writing studies, has reified into a rigid, linear pedagogical practice. This notion of process has been regularly questioned. In the 1980s, for example, Paul Kameen argued against notions of process based on cognitive, problem-solution models, arguing that once the problem is articulated, the solution is already contained in it. Invention, then, is only finding the solution that is already predetermined often by the teacher or the structure established via heuristics. He uses Coleridge's method as an alternative:

> Methodical thinking is dialectical in its operations—i.e., the specific route that inquiry will follow cannot be mapped a priori; it reveals its pattern as exploration proceeds, each step preparing the ground for its (often unanticipated) successor; and because methodical thinking is spontaneously self-questioning, it is more nearly subversive than recursive in its capacity to adjust to the unexpected. ("Coleridge").[9]

More recently, Thomas Kent's edited collection *Post-Process Theory: Beyond the Writing-Process Paradigm* takes a social-constructivist view against what has become a fairly simple, generalizable notion of process. In the introduction Kent counters, "writing constitutes a specific communicative interaction occurring among individuals at specific historical moments and in specific relations with others and with the world and that because these moments and relations change, no process can capture what writers do during these changing moments and within these changing relations" (1–2). While recognizing the importance of change and relations, most of the work in the collection seems to focus (explicitly or implicitly) on communication and the communications triangle, missing much of the complexity in an evolutionary model. Margaret Syverson is one of the few who have taken up complexity theory to counter a cognitive approach to the writing process and the simplicity of the communications triangle, substituting the static poles of writer, text, audience, and world with distribution, embodiment, emergence, and enaction (23). Such a remodeled theory of composing attempts to find detailed language and concepts that situate cognition in social and material development. Sounding somewhat like Taylor above, she notes that "a theory of composing as an ecological system is particularly vexing because it challenges our present investment in and assumptions about ownership of intellectual work, such as creative ideas and textual productions" (202). If everything co-evolves, drawing such distinctions as ownership becomes moot, an abstract game unrelated to an idea/text's emergence.[10]

Though Syverson is perhaps the best example in rhetoric and composition of trying to create an encounter with complexity theory, I'm worried that Syverson doesn't go far enough.[11] As she notes in her conclusion, there are many questions to be answered. If rhetoric and composition is to move forward and adapt to the coming networked cultures, it can no longer settle, much less strive for, the production of overly simple systems to account for the complexity of writing. The investigation of works like Taylor's, as I have tried to show, could push retheorizations of writing and rhetoric forward toward the coming global media culture.

NOTES

1. This article encapsulates the trajectory of my work over the past five years with rhetoric and vitalism, which argues for reassessing the concept of vitalism within a genealogy from Coleridge through Nietzsche, Bergson, Heidegger, Foucault, and Deleuze to complexity theory. As essentially an abstract of this much larger project, it is merely suggestive of the potential for this line of inquiry. Such a project could never be fully articulated in an article, perhaps not even in a book. Rather, it sets up nodes, fragments, sites to set the stage for future development. My current manuscript title for this project is "A Counter-History of Composition: Toward Methodologies of Complexity."

2. Berlin acknowledges an open-ended dialectic in much of his theory but in his practice he still over-values one element of the polarity and asks his students to do the same. For the sake of pedagogy, for the sake of action, he is willing to freeze the movement and make a valuation.

3. Biesecker writes, "If the subject is shifting and unstable, [. . .] then the rhetorical event may be seen as an incident that produces and reproduces the identities of subjects and constructs and reconstructs linkages between them. [. . .] [I]t marks their articulation of provisional identities and the construction of contingent relations that obtain between them" (126). In addition to writer and world, Biesecker goes on in her article to extend this argument to audience.

4. It should be noted that many of these key terms in complexity theory, as I am framing them here, are attempting to get beyond the basic, dialectical notion of the communications triangle that has grounded rhetoric and composition from Kinneavy to Berlin. It is not that the communications triangle is wrong, but that its simplicity can only take us so far.

5. The two concepts of *enkrateia*, or mastery, and *sophrosyne*, or moderation, form the beginnings of self-reflection. Foucault's insight is that these concerns over sexual activity "create the possibility of forming oneself as a subject" (138). It is this element in Greek thought that Christian thought follows and connects with Plato, setting the stage for an internal and eventually modern subject to emerge.

6. I am currently working on an article that examines the ethos/ethics of Foucault and Deleuze in further detail tentatively titled "On the Im/Possibility of Ethics: or, Toward an Ethico-Politics of De/composition."

7. See Rorty for a number of essays on Aristotle's theory of the emotions.

8. In *Spinoza: Practical Philosophy* Deleuze recognizes the problematic distinction of mind and body:

> It has been remarked that as a general rule the affection (*affectio*) is said directly of the body, while the affect (*affectus*) refers to the mind. But the real difference does not reside there. It is between the body's affection and idea, which involves the nature of the external body, and the affect, which involves an increase or decrease of the power of acting, for the body and the mind alike. (49)

9. See Kameen's "Rewording" in which he situates this argument within a Heideggerian framework in order to place it in the context of a more ecological development.

10. I stick to Taylor's use of the term evolution here, even though Deleuze and Massumi have gone beyond *evolution* to *movement*. In *Bergsonism*, Deleuze warns against two misconceptions in the use of the term *evolution*: seeing evolution as predetermined and seeing evolution as only occurring at the level of the actual, missing the level of the virtual, or affect (98–101). Massumi generally speaks of evolution positively—though he notes its connotation of progress, order, and predictability (218). The primary distinction for him is that evolution operates at the most global level (112); movement is what happens at particular nodes, its immediate relations that lead to larger scale change (evolutionary change at the level of the whole). Movement happens within particular bodies at

the level of affective relations. Massumi chooses to emphasize the term *movement* because the terms *process* and *evolution* are so over-coded that movement has been largely excluded in cultural theory (3).

11. For example, Syverson uses fairly standard research practices such as protocol analysis and ethnography. While I agree that some kind of research needs to be done on ecologies surrounding composition, my sense is that at some point we will need new research strategies more in line with the nature of complexity.

WORKS CITED

Barthes, Roland. *The Semiotic Challenge.* Trans. Richard Howard. Berkeley: U of California P, 1994.

Baumlin, James. Introduction. *Ethos: New Essays in Rhetorical and Critical Theory.* Ed. J. Baumlin and T. Baumlin. Dallas: Southern Methodist UP, 1994. xi-xxxi.

Berlin, James. *Rhetoric, Poetics, Cultures: Refiguring College English Studies.* Expanded Ed. West Lafayette: Parlor, 2003.

Biesecker, Barbara. "Rethinking the Rhetorical Situation from Within the Thematic of Differance." *Philosophy and Rhetoric* 22.2 (1989): 110-30.

Bitzer, Lloyd. "The Rhetorical Situation." Rpt. in *Rhetoric: Concepts, Definitions, Boundaries.* Ed. William A. Covino and David. Jolliffe. Boston: Allyn, 1995. 300-10.

Deleuze, Gilles. *Bergsonism.* New York: Zone, 1988.

——. *Spinoza: Practical Philosophy.* Trans. Robert Hurley. San Francisco: City Lights, 1988.

Enos, Richard, and Janice Lauer. "The Meaning of Heuristic in Aristotle's Rhetoric and Its Implications for Contemporary Rhetorical Theory." *A Rhetoric of Doing: Essays on Written Discourse in Honor of James L. Kinneavy.* Ed. S. Witte, N. Nakadate, and R. Cherry. Carbondale: Southern Illinois UP, 1992. 79-87.

Foucault, Michel. *The Use of Pleasure: The History of Sexuality* Vol. 2. New York: Vintage 1984.

Hawhee, Debra. "Kairotic Encounters." *Perspective on Rhetorical Invention*. Ed. Janet Atwill and Janice Lauer. Knoxville: U of Tennessee P, 2002. 16-35.

Kameen, Paul. "Coleridge: On Method." *Correspondences* 5 (Summer 1986): n.p.

——. "Rewording the Rhetoric of Composition." Rpt. in *Pre/Text: The First Decade*. Ed. Victor Vitanza. Pittsburgh: U of Pittsburgh P, 1993. 3-30.

Kent, Thomas. Introduction. *Post-Process Theory: Beyond the Writing-Process Paradigm*. Ed. Thomas. Kent. Carbondale: Southern Illinois UP, 1999. 1-6.

Kerford, G.B. *The Sophistic Movement*. New York: Cambridge UP, 1981.

Lanham, Richard. *The Electronic Word: Democracy, Technology, and the Arts*. Chicago: U Chicago P, 1993.

Massumi, Brian. *Parables for the Virtual: Movement, Affect, Sensation*. Durham: Duke UP, 2002.

——. *A User's Guide to Capitalism and Schizophrenia: Deviations from Deleuze and Guattari*. Cambridge: MIT P, 1992.

Miller, Carolyn. Foreword. *Rhetoric and Kairos: Essays in History, Theory, and Praxis*. Ed. P. Sipiora and J. Baumlin. Albany: State U of New York P, 2002. xi-xii.

Poulakos, John. *Sophistical Rhetoric in Classical Greece*. Columbia: U of South Carolina P, 1995.

Rorty, Amelie O., ed. *Essays on Aristotle's Rhetoric*. Berkeley: U California P, 1996.

Syverson, Margaret. *The Wealth of Reality: An Ecology of Composition*. Carbondale: Southern Illinois UP, 1999.

Taylor, Charles. "Heidegger, Language, and Ecology." *Philosophical Arguments*. Cambridge: Harvard UP, 1995. 100-26.

Taylor, Mark C. *The Moment of Complexity: Emerging Network Culture*. Chicago: Chicago UP, 2001.

Ulmer, Gregory. *Heuretics: The Logic of Invention*. Baltimore: Johns Hopkins U P, 1994.

Vatz, Richard. "The Myth of the Rhetorical Situation." Rpt. in *Rhetoric: Concepts, Definitions, Boundaries.* Ed. William A. Covino and David Jolliffe. Boston: Allyn, 1995. 461–67.

Vitanza, Victor. *Negation, Subjectivity, and the History of Rhetoric.* Albany: State U of New York P, 1997.

PART THREE

Network Culture

8
GRRRL ZINE NETWORKS

RE-COMPOSING SPACES OF
AUTHORITY, GENDER, AND CULTURE

Michelle Comstock

> [Zines] are little publications filled with rantings of high weirdness and exploding with chaotic design.... In zines, everyday oddballs [are] speaking plainly about themselves and our society with an honest sincerity, a revealing intimacy, and a healthy "fuck you" to sanctioned authority—for no money and no recognition, writing for an audience of like-minded misfits.
>
> —Stephen Duncombe

> From the Revolutionary rabble-rousing of Thomas Paine's pamphlets, to the Unabomber manifesto, Do-It-Yourself publication is as American as Mom's homemade pipe bomb.
>
> —Ann Magnuson

A wide range of workplace and civic writing communities have garnered the attention of many composition scholars and teachers, yet the everyday literate activities of a growing do-it-yourself (DIY) zine network have remained relatively unaddressed in the various realms of academic inquiry.[1] There may be many reasons for this omission, not the least of which is the decidedly nonacademic layout, style, grammar, tone, and content of the zine genre as well as its history in American counterculture. However, if we as teachers and researchers of writing wish to continue situating academic literacies within the larger framework of cultural production and pedagogy, the zine movement and its most recent progeny—what is known as the riot grrrl zine scene—have much to teach us about the sites, practices, politics, and economies of writing. As this article will demonstrate, grrrl zine editors are collectively engaged in forms of writing and writing instruction that challenge both dominant notions of the author as an individualized, bodiless space and notions of feminism as primarily an adult political project.[2]

By appropriating the political tactics and writing practices of both the punk zine scene and the larger feminist movement, the grrrl zine network is creating new spaces for postfeminist authorship—a term I use cautiously given its popular "post-revolutionary" and anti-feminist connotations. As I will later explain, these new authorial positions (and positions of authority) are not necessarily confined to biological age or even to a particular generation but are informed aesthetically and politically by the historical events and ideological movements of the late 1980s and 1990s. Postfeminism, as I use the term here, does not entail a rejection of second-wave feminism, nor does it foreclose the promises of this earlier movement; instead, it represents an adjustment by grrrl zinesters of these earlier feminist principles, including the very privileging of gender as an universal category. As Ednie Garrison states in her study of grrrl (sub)cultures, "The shift from speaking about 'women' as a unified subject to a recognition that women are not all the same, nor should they be, is something most feminists, young and not as young, take for granted in the 1990s" (145). This recognition is not simply a generational or recent phenomenon, but builds on longstanding efforts in feminism to acknowledge racial, class, sexual, and cultural difference.

Situating the grrrl author as a postfeminist author politicizes and historicizes her newly adopted stances, styles, technologies, and practices of writing. We may, therefore, look to this contemporary grrrl zine network for more than just new sites of authorship; we may also understand these young and not so young women as rhetoricians engaged in the important political processes of re-envisioning and revising "feminism" and "girlhood" in the contemporary United States. Before discussing some of the more specific transgressive or postfeminist writing sites, practices, and economies, however, I want to briefly situate the grrrl zine's formation within the context of grassroots, countercultural publication, a context that is often at odds with larger, more mainstream ideological and political movements such as feminism.

ZINE SCENE: A SITE OF EXTRACURRICULAR WRITING

With its roots in the small press ("lilmag") and fan magazine communities of the 1950s and 1960s, the contemporary zine scene comprises a significant extracurricular site of reading and writing pedagogy, where editors, writers, and readers learn the critical practices of countercultural production and distribution outside of formal institutional settings. Composed of "rants" against the homogenizing effects of mass culture and popular media, zines forego the grammar, layout, content, and distribution methods of conventional publication. Most of them originally appeared on "one sheet of legal-sized paper copied on both sides and folded three times, trimmed and stapled to make a 16-page 'mini' zine" (Vale 4).

According to Stephen Duncombe, the typical zine begins with a "personalized editorial," moves to "a couple of opinionated essays or 'rants,'" and concludes with reviews. Poems, interviews, drawings, comics, stories, solicited letters from friends, and reprints from the mass media are also scattered throughout (Duncombe 10). Zinesters are notorious for mixing genres and strategically combining personal stories, fiction, rants, poetry, and essays, which are practices that are facilitated by photocopiers and cut-and-paste desktop publishing programs (see Green and Taormino xi). Undoubtedly, the zine tradition has undergone a great deal of "mainstreaming" in the last few decades, with slick cyberzines now dedicated to the arts of browsing and buying online products. However, this so-called co-optation of the zine genre seems to have occurred just as more women and girls became involved in online and offline zine writing.

In critical response to the male-dominated punk zine scenes of the 1960s, 1970s, and 1980s, the grrrl zine movement has constructed a space for young women to act as writers, designers, artists, and Web designers. They have challenged not only the gendered hierarchies of alternative writing cultures, but also the exclusionary sites and practices of mainstream authorship. In the late 1980s and early 1990s, the riot grrrl "rants" against dominant notions and images of girlhood and femininity first appeared in the punk and grunge music subcultures of Olympia, Washington and Washington D.C. As Emma of *Riot Grrrl 5* explains, "Riot Grrrl is about *not* being the girlfriend of the band and *not* being the daughter of the feminist. We're tired of being written out—out of history, out of the 'scene,' out of our bodies. . . . For this reason we have created our zine and scene" (qtd. in Duncombe 66). Many grrrl zinesters, such as Mimi Nguyen, began their writing careers in response to sexism in the punk music and zine scene. In a widely reprinted editorial written for *MaximumRocknRoll*, Nguyen argues that a column by Rev. Norb on Asian women's sexuality is a "racist fetishization of a fragmented female body" and is "ideologically comparable to the imperial European fascination with the 'Hottentot Venus.'" Nguyen's editorial, along with a series of other letters written by women musicians, inspired a string of stories (and later entire zines) about the role of women in the punk movement.

Appropriating the term "girl"—much in the same way punks had appropriated the term "loser"—riot grrrl, with its triple *r* growl, redeploys the term to serve both as a critique of dominant and punk girl images and as an alternative collective identity for young women writers. The following riot grrrl "call-out" below testifies to the use of writing as a tactic for critical or alternative subject formation outside both the male punk scene and the mainstream feminist movement. Here, zinesters create a radical grrrl self without rules, labels, or limitations, a grrrl who "needs to make some noise":

GRRRL PoWeR = GRRRL LoVe =
ReSpEcT = EnCoUrAgEmEnT =
StRoNg SeLf ImAgE =
DeSiRe tO TeAr DoWn tHe RuLeS =
GRRRL PoWeR =
R E V O L U T I O N.

This is a page for the grrrl in every female out there who needs to MaKe SoMe NoIsE!! We are here to listen to one another and define the meaning
of "Girl Power" and I'm not talkin' about the Spice Girls!!

GRRRL PoWeR iS:

feeling okay about being a girl: Be proud! We ROCK!

promoting girl love and friendship: A kind of sisterhood. Don't talk to me about cliques or sororities; in this clique there are no rules, no certain way to be, and we don't leave anyone out!

encouraging one another: Telling each other it's cool to be who they are and let them express themselves!

teaching: girls, boys, men, women, old or young about grrrl issues things that affect each one of us (equality, individualization, the right to speak your mind and let your thoughts run free).

respecting each other: to realize the individuality of every girl on this planet, not to divide people into groups like race, religion, ethics, etc., to look down upon derogatory names and phrases against girls and anyone else. ("Riot Grrrls")

As this manifesto indicates, the postfeminist revolution is founded on difference, individuality, and a continuing effort to define the loose cultural and political spaces of grrrl power and sisterhood. Although riot grrrl manifestoes like this one were originally distributed via grassroots networks (for example, independent bookstores, reading groups, rock concerts, and community centers), many are now created and dispensed online for a larger, more diffuse mass audience. Rhetorically speaking, the grrrl zine movement—with its changing spaces, purposes, imagined audiences, contents, and political effects—offers us a glimpse at the shifting contours of contemporary authorship and countercultural production. While compositionists have written astutely about the zine genre's promise for noncanonical reading and writing in the classroom, few have seriously explored what zines, particularly riot grrrl zines, tell us rhetorically about acts of writing in complex political and cultural contexts.[3]

As writers and designers working collectively to counter dominant cultural images of adolescence and girlhood, grrrl zine culture comprises yet another significant site of what may be called "extracurricular" writing instruction, defined by Anne Ruggles Gere in another context as a place where writers learn how to produce and distribute texts outside of formal educational settings (see "Kitchen" 80). Like the writing groups documented by Gere in *Intimate Practices*, grrrl zine networks are gendered sites of cultural production and pedagogy in critical relation to more bureaucratized high school and college writing classrooms. The impulse to inscribe an alternative girl self—or rather an alternative girl space outside more formal writing environments (where one can "say what you want to say and not be afraid")—testifies to the importance of writing as a site of (counter)cultural pedagogy. As Linda Brodkey argues, it is this highly rhetorical "turn" to writing that we wish to replicate in our classrooms (140).

In what follows, I provide a rhetorical mapping of grrrl zines as sites of cultural production and pedagogy, where the very spaces, practices, economies, and politics of (feminist) authorship are in dispute. Although necessarily brief and partial, this mapping highlights the extent to which zine writers and Web designers have reterritorialized grrrl identities as sources of political and social authorship and agency. The grrrl zine scene has offered many young women ways to reconceive their participation in the public sphere not just as consumers but also as producers and writers of culture. Furthermore, the grrrl zine has also been a means for young women to take up authorial positions outside the academy's often paternalistic discourses of student need and "banking" approaches to literacy education.[4]

In many writing classrooms, we have already begun to recognize how traditional notions of literacy, authorship, and the text marginalize viable alternative forms of writing and communication. Bruce Horner claims that imagining authorship as a position occupied by individuals distinct from social and historical pressures can lead us to deem certain kinds of literacies "marginal" (505). Dominant notions of what constitutes work, writing, and authorship, according to Horner, "blind us (and others) to the kind of cultural work . . . accomplished in the writing and reading in which we and our students engage" (522). Aware of the important political and cultural stakes, many writing teachers place less focus on texts and products and grant more attention to the social conditions and practices that mark the space of "authorship." The space of the grrrl author, for example, is marked by the rhetorical desire to network young women who are isolated in homes, high schools, and colleges, and provide them with alternative literacies (tools and knowledges) for self-expression and solidarity.

As the following discussion demonstrates, the struggle over grrrl authorship is an embodied one that is often articulated at the site of the traumatized, adolescent female body. This rearticulation, like most forms of writing, is not without its significant economic and political effects. Especially as more young women choose to publish their zines in online environments, the lines between corporate sponsorship, popular culture, and (post)feminist political action have become increasingly difficult to trace. Whether it is in their initial encroachments on the boy-centric punk zine scene or in their infiltration of a mostly male World Wide Web, grrrl writers and designers teach us that authorship is not a fixed or completely predetermined category but a site of collective struggle and interactivity.

THE GRRRL WRITER AND EMBODIED AUTHORSHIP

> It's amazing that all these young (mostly white) women have decided to redesign the whole world according to the architecture of their private-made-public traumas and promises of "girl love" wish-fulfillment. Riot Grrrl is amazing in so many ways: as confrontation, as education, as performance, as aesthetic, as support, as theory, as practice, etc.
> —Mimi Nguyen

> You don't need to be a punk.
> You don't need our permission.
> There are no rules.
> No leader.
> Every girl is a riot grrrl.
> —"Riot Grrrls"

The "grrrls want to riot too" slogans of the early riot grrrl zine movement opened up room for many young women writers (mostly white and middle class) in the male-dominated world of DIY publication. Building on the larger zine scene's ethos of shared difference and communitarian anarchy, grrrl editors began composing an embodied, postfeminist collective identity grounded in the "private" traumas of rape, incest, and eating disorders, as well as the "public" issues of job discrimination, sexual harassment, and reproductive rights. Almost every riot grrrl zine in the early 1990s contained a poem or story about sexual abuse by a male friend or family member; rants about self-image, including the mass media images of women; and manifestoes explaining what a riot grrrl is and her place in the world.

According to Duncombe, the process of "sharing their stories with [each] other and pointing their fingers at the accused" allowed these young women to "express their rage, relieve their shame, and overcome the isolation that accompanies such an experience" (67). In turning to zine writing and publishing, however, grrrls not only expressed rage, they also radically altered the places and subjects of authorship, resituating them on the conflicted discursive and material sites of the traumatized girl body. The grrrl body, the girl writer, like any other authorial position, is a site of gender, racial, and sexual struggle. In the second issue of her famous riot grrrl zine, *Bikini Kill,* Kathleen Hanna describes the grrrl writer and performer as a grotesque mixture of gender and sexual contradiction:

> to be a stripper who is also a feminist, to be an abused child holding a microphone screaming all those things that were promised, in one way or another, "I won't tell." these are contradictions I have lived. they exist, these contradictions cuz I exist. . . . because I live in a world that hates women and I am one . . . who is struggling desperately not to hate myself and my best girlfriends, my whole life is constantly felt by me as a contradiction. in order for me to exist I must believe that two contradictory things can exist in the same space. this is not a choice I make, it just is.

In Hanna's case, the traumatized grrrl body, with its contradictory and tabooed functions, becomes a performative site of writing and collective space-making. The scenes of authorship are displaced onto "out of place" bodies that bulge, bleed, and ooze over the boundaries of an appropriate and normalized white, middle-class femininity and constitute a community of young women mutually appreciative of a contradictory body out of control. Historically, girl groups have challenged the patriarchal control of public space, along with its privileged authorial positions, by articulating and performing a grotesque female body and its disharmonious representations (for example, its very physicality and sexuality) that mark its territory. This includes affirming through narratives and images of bodily behavior and functions deemed private and threatening to the "disembodied" realms of male-dominated publics and cultures (see Campbell 16).

The difficulties associated with rearticulating the grrrl body as a site for pleasure and resistance are the subject of many zine columns and rants. For example, Tammy Rae Carland writes, "I've been thinking about how I think about my body. Or rather how I don't think about my body. About how I was taught not to consider my body as a site for pleasure and/or resistance. It was the place where I stored memory and secrets, and it was the thing that attracted unwanted attention. In other words, it was an awkward container" (46). Carland's observations suggest the political and social stakes in appropriating the adolescent, female body as a forum for expression and solidarity. School literacies rarely provide the tools and knowledges for constructing this type of forum, especially as it relates to violence and sexuality. In high school and university writing courses, for example, many women are met with fear or confusion when they attempt to articulate their experiences of bodily trauma. For example, Michelle Payne discusses the anxieties that teachers experience when confronted with "personal" student essays on violence in relation to the female body. This kind of writing, argues Payne, "challenges our purposes, our roles, and our power as teachers. . . . Subjects such as abuse, suicide, death, and divorce are perceived as more closely connected to a private, more vulnerable sense of self, a self that some believe does not belong in a writing class" (xvii-xviii). Although many grrrl editors have spent many hours writing and reading in university classrooms, they have chosen the zine scene and not the academy as their vehicle for change. For them, the grrrl body is a textual body (an authorial position) marginalized by the dominant discourses of the university. As Payne claims, this marginalization is partly due to continued cultural anxiety about the female body as well as by adult anxiety about the adolescent body, which is often translated in terms of disruption, trauma, change, and violence.

Playing on this anxiety, writers such as Hanna irreverently invert the common understanding of white, middle-class female adolescence in order to create an alternative grrrl aesthetic and politic. The sexually knowledgeable, "out of control" female adolescent asserted by *Bikini Kill* and other grrrl zines appropriates and deconstructs the ubiquitous virgin/whore figure of adolescence, producing what has been described as the "kitten-with-whip grrrl aesthetic" (Burana 76). This grrrl aesthetic draws on two powerful images of cultural consumption—the hyperfeminine "Hello Kitty" iconography associated with the tools of school literacy (perfumed erasers, pencils, backpacks, and carrying cases) and the hypersexual, dominatrix iconography associated with whips, weapons, and the leather-clad Tank Girl. Refiguring these images allows zinesters to construct a new aesthetic and a new forum for grrrl solidarity that together challenge the gendered power relations associated with zine writing and (counter)cultural production. The kitten-with-whip aesthetic and ethos, however, do not work to desexualize the grrrl writer and her audiences; rather, they seek to create what Mary Ann Doane calls in a different context a "terrain of fantasy" where oppressive representations of girls and women are playfully denaturalized (180). In many ways, writing and reading grrrl zines is another instance of feminist camp or gender parody, defined by Pamela Robertson as "a critical tool and a promising means of initiating change in sex and gender roles" (10). Through text and images, grrrl editors playfully make stylized femininity fantastic and literally incredible. The grotesque morphing of the cute, mouthless Hello Kitty style with a screaming punk rock and S/M iconography demonstrates the reader/user's recognition and misrecognition of herself in these stereotypical images (see *Chickclick*; *Hellfire*; and Chan, "Riot"). The grrrl writer's conflicted kitten-with-whip identity is informed not only by contradictory notions of gender and adolescence, but also by complicated and intersecting discourses of class, race, and nationality.

Grrrl solidarity is an exclusionary project, operated by young women with the necessary leisure time and resources. And, as previously stated, it is a project that centers on a primarily white, middle-class femininity. Nguyen, creator of *Slant*, rants against the network's racism: "As a resident racial agitator, I'm cast in the role of an 'outsider' rocking the boat. Past discussions of race issues have been cut short as girls and women bemoaned to 'fighting' as bad for grrrl unity, meaning, racial difference is bad for grrrl unity" ("White"). Although Nguyen and others construct a mostly white zine network, in many instances grrrl editors identify as "mixed race" or "women of color" and highlight those contradictions that inform overlapping racial, national, and sexual identities. Sabrina Sandata of *Bamboo Girl* began her zine in early 1995 because she "couldn't find anything to read on girls like me, a mutt (Filipina/Spanish/Irish/Scottish/a little Chinese) who was in-your-face about issues within the hardcore/punk and/or queer communities." For Sandata, the turn to writing involved a movement toward those things that were "'not nice for girls to be thinking,' especially for 'a nice little *Asian* girl'" (98). More than just countering mainstream white images of girls and women, however,

Sandata's and Nguyen's work highlights the fragmented nature of American "girlhood" and underscores the idea that the girl self is always under construction and is resistant to easily identifiable gender and race categories.

Of course, not everyone can easily comprehend the fragmented identities of the grrrl zine movement. As I suggested earlier, its imagined and intended audiences are not necessarily academic nor even political. In fact, most of the rants, letters, and reviews seem rather solipsistic. Zinester Leah Lilith Albrecht-Samarasinha's use of a personalized confessional discourse is common: "In *Sticks and Stones*, I published the rants about abuse in my family that I'd been writing in my journal since I'd got the hell out, but had never even shown my close friends" (3). For Albrecht-Samarasinha, zines represent "homemade productions from other crazy girls that told me I was not crazy or horrible, a freak isolated from all the world" (3). The effects of this "homemade" network, however, have been social as well as personal, as many young women have begun to understand themselves as writers and participants in a significant, alternative public sphere. Similar to Nancy Fraser's concept of "subaltern counterpublics," zines such as *Bikini Kill* serve a dual private and public function. The zine's rant spaces, on the one hand, offers places away from oppressive home, school, and work environments where young women can confess, receive support, and regroup (Fraser 123). On the other hand, they constitute a "training ground" for cultural and political activities directed toward wider publics (see Fraser 114). The dialectic between these two functions—private enclave and public training ground—gives these grrrl counterpublics their emancipatory potential within the stratified societies of both mass and alternative cultures.

THE GRRRL ZINE SCENE AND NETWORKED AUTHORSHIP

Piecing together the grrrl counterpublic is less an act of authorship in the conventional sense than an act of critical editorship. Grrrl zinesters selectively cut and paste the styles and genres of popular and alternative cultures and engage in writing as an intrinsically networked process of both consumption and production. The assumption of any writing subject position entails inheriting a whole set of social and economic practices and conditions that define the very parameters of "authorship" in that community. Likewise, becoming a grrrl writer requires the acquisition and internalization of the practices and codes of the "private-made-public" grrrl network. These practices include methods of critical pastiche inherited from the world of countercultural zine publishing, along with forms of political and cultural resistance borrowed from the more established feminist movements.

Committed in part to punk principles of anarchy, the grrrl movement tapped into the grassroots production and distribution practices of a larger zine and amateur press scene that historically have resisted mainstream and academic notions of authorship as ownership. With lengthy "reader letters" sections and magazine cut-outs, grrrl zines often work to confound easy reader/writer and copyright distinctions. And although many may develop a strong sense of individual voice, most grrrl writers use the resources of typeface, layout, and writing style to demonstrate multiple contributions. Furthermore, extensive attention to letters and reviews makes grrrl zine writing less about individual creative expression (though there is emphasis on that) and more about providing a network or forum for writers, artists, and musicians on the fringes of mainstream culture. Zine production and circulation practices (for example, cutting and pasting found media images, handwriting rants and manifestoes, photocopying, folding, stapling, and mailing with reusable stamps) follow what Michel de Certeau calls a "re-use" or "reappropriation" of mainstream marketing structures. With the decline in cost of personal computers and the influx of desktop publishing tools into schools and workplaces, for example, many grrrl zinesters "liberate" computer and copier time for their own countercultural work. As de Certeau argues in another context, they attempt to "deflect the technocratic functioning of the mainstream and popular presses by means of a multitude of tactics" and employ a rhetorically savvy, context-specific, makeshift creativity associated with "groups caught in the nets of discipline" (xv).

Makeshift grrrl writing styles and practices revel in anti-discipline and the improper. Zinester Pagan Kennedy's approach is typical: "Almost instinctively I broke every rule of respectable fiction. I published my own work (for serious literary types, self-publishing is considered a sign of rank amateurism). I drew pictures. I wrote unpolished sentences and hardly went back to revise" (9). This impropriety stands in direct opposition to the ethics and values of another writing scene—the university classroom. While university writing classrooms often become what Brodkey calls "middle-class holding pens," where students learn to write well-formed and well-mannered essays displaying their "knowledge of and fealty to middle-class values," the grrrl network places value and worth on resistance to these middle-class values, especially where they concern the reproduction of white, middle-class femininity (135). Writing for many of these young women demands the reappropriation of the tools and practices of school literacy in order to highlight codes of femininity. According to Amanda, cofounder of a West Coast teenage feminist collective, the value of grrrl zines is their ability to break the rules of both conventional essay writing and traditional femininity: "There are so many things that we [young women] are not allowed to say; if we only just fucking screamed them and made people hear them, then it would become okay." Zinester Toad explains that ranting is "taking the ugliest detail and amplifying it 500 times—that way everybody gets used to it" (see Vale 48).

Like its style and tone, the distribution methods associated with grrrl zine publishing are networked, rhetorical practices that form an integral part of the overall writing process. Circulation counts, for example, are scaled to maintain a sense of "personal contact" with readers, a strategy illustrating the high value zinesters place on social capital over economic profit (Duncombe 12). Slow production time and erratic release dates also illustrate that the primary reward for writing is not necessarily monetary profit (though economics are important), but entry into the world of misfits and independent grrrl publishers. Within the zine scene, as in most literary writing environments, there is what Pierre Bourdieu calls a homologous connection between symbolic and material capital. Similar to the avant-garde publisher profiled by Bourdieu, the grrrl zine editor insists on more direct contact with writers, readers, and the work; thus, she is able to confront the financial risks "by investing ... in undertakings which can, at best, bring only symbolic profits." As Bourdieu points out in another context, an editor makes this investment, however, "only on condition that he [or she] fully recognizes the specific stakes of the field of production and ... pursues the sole specific profit awarded by the field, at least in the short term, i.e. 'renown' and the corresponding 'intellectual authority'" (100). In the case of the contemporary grrrl zine scene, this symbolic capital or "renown" comes in the form of citation (inclusion in other zines and catalogues) and a wide but controlled, hand-to-hand distribution. All of this works toward the final "pay-off": membership in the network.

Membership in the grrrl network depends partly on the articulation and enactment of what zinesters call "girl love," the principles of which are passed on in lists and manifestoes intended to unify as well as exclude:

> Girl Love is ... making space where women/girls feel unthreatened and unintimidated ... talking about abuse and rape when no one else will listen ... learning and teaching each other how to do stuff and be active ... screaming in public ... knowing that you are connected to all girls and the way you view yourself is related to their self-image as well ... sharing resources with other girls ... helping each other see our beauty and build our own culture around what we see ... refusing to let companies prey on our insecurities in order to get our money ... trying to understand how oppression and the status quo work and how we fit into it ... reclaiming our customs and rituals (hanging out in the bathroom, slumber parties, shopping, the color pink, whatever we fucking want). (Chan, "Girl")

In a society where cultural and economic capital depends on "networks of contacts and the information, norms, and values filtered through them," these grrrl writers seem cognizant of the economic and symbolic power of reclaiming the exclusive rituals of middle-class, suburban girlhood (Hanson and Pratt 7).

The cultural hierarchies associated with these rituals are expressed in the following passage from the grrrl zine *doris*. Here Cindy O. demonstrates a recognition of the economies of writing and distribution and their relationship to the codes and practices of "girl love":

> i was on the bart train the other day and there were these four girls, high school girls. three of them looked just like TV, long hair that they would brush back with their fingers. all four had practiced facial expressions, small noses, lines drawn around their eyes. but one girl, she was too tall and gangly and it looked like she just got her

> braces off, the way she kept feeling her teeth with her tongue. her backpack had paintings of suns and moons and flowers that you could tell she painted on there, and you could tell her friends made fun of it behind her back. she was the one i watched. her backpack was unzipped part way and i snuck doris number one in, number one, full of my secrets. i couldn't hand it to her because i knew she wouldn't take it, not with her friends watching. i snuck it into her life for her to find later, alone, in her bedroom. the people on the train who saw me do it glared at me, mean and suspicious, like i'd stolen something from that girl, and maybe i had. i got off at the next stop. (71)

The passage illustrates, among other things, how methods of grrrl zine distribution depend on the tracing of young women's routes through the city and the profiling of those who look like they belong. This kind of distribution or exchange, as Cindy O. indicates, comes with its own set of cultural and economic gains and losses. Writing and reading grrrl zines implies travel into new forms of affiliation, as well as exclusion, and the adoption of values and beliefs resistant to yet dependent on the TV world of mainstream girlhood.

These forms of affiliation have been further complicated by the grrrls' movement online. Although the Web has allowed many more young women access to network literacies, the relatively anti-commercial grrrl zine scene must now contend with the corporate distribution and marketing structures associated with Web writing environments. Internet and Web technologies are changing the practices and economies of most writing contexts, and the online grrrl network is no exception. The struggles over online grrrl authorship and sponsorship illustrate many of the changes in larger, more established writing communities inside and outside the academy.

ONLINE GRRRL WRITING NETWORKS: THE PROBLEMATICS OF SPONSORSHIP

When grrrl zinesters became Web-mistresses in the middle to late 1990s, the "revolution grrrlie-style" rhetorics of the print zine movement acquired different forums and formats in their shift to online environments. Now mediated by high-speed computer networks, the online grrrl zine scene highlights many of the limitations and promises of hypertext writing and Web contexts for progressive political practice. Compositionists have celebrated the disruptive potential of hypertext writing, claiming that hypertext's multivocality and nonlinear format challenges traditional academic discourse, while many researchers have also noted its failure to pose truly radical textual interventions.[5] In their article on the relationship between hypertext theory and feminist praxis, Donna LeCourt and

Luann Barnes reassess the ability of hypertext to enact a feminist textual intervention into the conventional, logocentric discourses of the academy. Their conclusion emphasizes hypertext's potential for "speaking the multiple subject," while at the same time noting its inability to transcend particular discursive contexts (66). Although their focus is primarily on more localized, academic contexts, LeCourt and Barnes' critique could also extend to the larger technological and commercial contexts of hypertext writing. Indeed, as many grrrl zine editors have discovered, writing on the Web is necessarily embedded in the discursive and material contexts of corporate sponsorship.

The globalized production and distribution technologies of Web writing have challenged the DIY and coalition-building practices of the riot grrrl movement and diffused their ongoing efforts to build separate grrrl spaces and readerships. The context of the Web is, as we know, a place where corporations increasingly mediate the production, distribution, and reception of even the most radical forms of critical literacy. And although online zines still hold promise for young women writers and artists as sites for social critique and change, it is becoming increasingly difficult to discern who exactly sponsors or regulates these newly formed sites of critical literacy.

A simple Web search for grrrl zines and groups, for example, reveals an ongoing dispute among grrrl editors, Web development companies, and media corporations over the virtual and material domains of grrrl power. Most search engines will locate site categories relating to riot grrrl conventions and feminist groups. While many sites situate grrrl identities and practices within the more anti-commercial, not-for-profit riot grrrl zine movement, other sites linked to the "grrrl" keyword construct a vastly different discursive terrain. If you type in "grrl," categories referring to women and computing, shopping, sports, and e-zine production appear. One such site, *Planet Grrl*, claims,

> We try to encourage, support, befriend, teach, respect and recognise other women and grrls online regardless of race, religion, political standpoint, age, orientation, employment status or anything else. Promote individuality and freedom of expression and opinion through new media. Have a sense of humour and laugh at ourselves and the rest of the world. Enjoy being a grrl. Have some fun!!
> . . . We're called PlanetGrrl because it was one of the few grrl domains not taken by a porn site or the ChickClick gals. It means Nuthin. Really. Except it's for Grrls all over this planet.

Unlike their print zine colleagues, Cybergrrl Alison Gianotto and Riotgrrl Nikki Giovanni seem more committed to making a name in the commercial end of technology as professional writers, Web designers, or computer technicians. This becomes more possible as the grrl signature—with its tough, edgy connotations— assumes cultural currency in these sectors. "It's interesting how the concept of 'grrrl/grrl' has morphed and changed," notes *Bikini Kill* contributor, Kristy Chan. She continues, "There are a lot of places most feminist women I know wouldn't want to be associated with—you know, the sites that are like Cosmopolitan girls in tough-girl gear yet still conveniently kissing the Man's ass—but many others that offer differing views that have enlightened and broadened the meaning of being a young and powerful woman" ("Riot"). Although the number of r's may signal different writer identities and political agendas, in the context of the Web, the placement of the r's also indicates ownership over domain space, thus making the connections between authorship and economic capital more explicit. Resembling the trademarks or copyrights of print culture, domain names such as "riotgrrrl" are for sale. Once claimed or bought, the domain becomes a restricted area, leading to the influx of "girl" derivatives listed above.

Geekgirl zine editor Rosie Cross, for example, fought to no avail in 1997 to protect her trademarked grrrl territory from Web entrepreneurs, who, according to Cross, were capitalizing on her zine's popularity (Salkowski). More likely, they are capitalizing on the wider significance of the geekgirl writer identity. Entrepreneur Helana Peterson took on the name "geekgrrl" because she was looking for something "aggressive." She explains, "The industry I'm involved [with] is male dominated, and I didn't want to have some flippidy foo name." To counter these attempts at appropriation, Cross called on the part of the DIY zine ethic that privileges originality: "I have a trademark, reputation and have spent considerable money protecting my name, identity and integrity. . . . The world knows I am the original, the only, one true geekgirl" (see Salkowski).

Joe Salkowski's "Geekgirl Grudge Match" highlights a more general debate over the sites of authorship on the Internet. Authoring a protected or resistant space of one's own has become problematic not only in terms of domain ownership but also in relation to the very interactivity of online writing environments. The complexity and mutability of Web-page production and reception processes, in which authorship is established through inter-networked travel, makes it impossible to trace unmediated lines of grrrl or feminist resistance. *Geekgirl*, for example, is a highly interactive site; as such it dissolves the traditional boundaries of authorship and ownership associated with writing and design. Following the mixed formats of the print zine genre, many *geekgirl* articles do not carry bylines, just Web addresses, and the ones that do cite authorship are recycled from books, Web sites, and print zines. In fact, most online zine articles are light on content and heavy on links or directories that serve to mark online networks. Thus, they often resemble what Jay Bolter has called, in another context, "topographic" writing, a "writing *with* places" as distinct from the writing of a place (25).

Buried under sporadic editorial content, however, are indications of *geekgirl's* growing connections to institutional lines of distribution or sponsorship. More mainstream "adult" organizations are recognizing and appropriating *geekgirl* as both a cultural and pedagogical site. The New Museum of Contemporary Art in New York, for example, featured *geekgirl* in its "alt.youth.media" exhibition, which explored video and media works intended for youth audiences. The Museum of Victoria in Australia also integrated the site into its traveling exhibition on Internet literacy called "cyberzone," which allowed thousands of "unconnected" people (mainly youth) to access *geekgirl's* resources. As people log onto *geekgirl* from various public and private sites and contexts, the lines between cultural and economic capital begin to blur. How, for example, are the patrons at the New Museum of Contemporary Art reading or viewing *geekgirl*? As part of a larger political project or as an individual design portfolio or both? *Geekgirl's* acknowledgment pages identify funds from organizations as diverse as the Australian Film Commission, the University of South Wales, and "the grrrls at Microsoft." It therefore becomes impracticable to identify who or what at any given time *geekgirl* is mediating. Are the articles and images mediating between grrrls and the Internet "boy's club"? Or between nonprofit government agencies and their technologically illiterate citizens? Questions such as these—questions regarding the political and cultural agency of writers and writing communities—are symptomatic of late capitalist economies, where the boundaries between micro-media and mass media and global industries and local businesses are in flux. Thus, distinctions between grassroots writing communities and more established institutional and commercial ones now tend to obscure more than they explain.

Not all grrrl editors have been comfortable with the commercial (or for that matter interactive) readership of Web writing space. Their discomfort resembles that of many writers first confronted with the virtual audiences of Web address. Some have taken their zines offline, preferring the more controllable, slower rhythms of print authorship. While others, like lisa, rarely update their pages, claiming their first priority is to their more local print audience: "i'm considering taking my webpages down soon. i'm feeling really anti-internet lately and too vulnerable. i get about 50 hits a day now and it scares me that that many people are reading this. plus i've become too dependent on my webpage and it's taking me away from actual paper zines." Echoing the concerns of many of her leftist colleagues, *Cupsize* editor Emelye claims that Web writing facilitates a dangerous dependence on the computer market and its chosen experts. Again, these critiques signal an understandable desire to control the spaces of authorship and audience and to root zine authorship and agency within a unified, physical sense of self and experience.

I do not wish to imply that print zine authorship is any less fraught with struggles over control, distribution, and ownership. Indeed, as more grrrl writers have entered the zones of mainstream print publishing, they have contended with restrictions not yet firmly established in online writing environments.[o] Internet authorship, more than mainstream print authorship, seems amenable to the DIY ethic of the print zine movement. Those privileged enough to have access to a computer, modem, Internet software, an e-mail account, and some Web design knowledge can publish online. This access is complicated, however, as disputes over the boundaries of grrrl text space continue and more e-zines editors are forced to post corporate banners or logos. And even as some of these sites strive to be "grrlie"—storing Web resources and tools for grrrl users—many also seek to be as "professional" as possible with state of the art Web designs and navigational tools. Although issues of corporate sponsorship and co-optation hang over grrrl e-zine projects, many editors—such as Cross, Chan, and Nguyen—are attempting to compose diverse, imagined communities of young women with multiple and asymmetrical points of access and identities. This kind of writing can constitute a significant form of cultural production and resistance in a moment when social and economic power is inextricably related to the knowledge and use of communication technologies.

GRRRL ZINE WRITING AND THE FUTURE OF (POST)FEMINIST AUTHORSHIP

The grrrl zine scene's ambivalent relationship to both the corporate mainstream and the countercultural raises questions about the shifting spaces of political or radical authorship. Are traditional notions of radical authorship indeed possible in today's mass-mediated writing communities and environments? In the case of the grrrl zine network, such questions too often degenerate into divisive, generational debates over the political efficacy of "girl power" or DIY feminism. Many in the mainstream and feminist media have (mis)read the DIY grrrl ethic as an attempt to sell watered-down versions of feminism to white, middle-class, high-school and college-aged women. The grrrl zine scene, according to feminists and antifeminists alike, constitutes just another cultural fad, or worse, a highly commercial, reified site of individualism. In a recent issue of *Time* magazine titled "Is Feminism Dead?" reporter Ginia Bellafante asks, "Want to know what today's chic young feminist thinkers care about? Their bodies! Themselves! . . . In the '70s, feminism produced a pop culture that was intellectually provocative. . . . Today it's a whole lot of stylish fluff. . . . The voice of the movement, *Ms.* magazine dissected women's roles, status and pay. [Now the] voice of the 'new girl order,' *Bust* sarcastically dissects often wacky sexual exploits" (54–56). In a patronizing tone, Bellafante attributes the "new girl order" to a narcissistic obsession with one's body and associates the new postfeminist rhetorics with the apolitical discourses of self-help manuals. Debbie Stoller, coeditor of *Bust*, contested Bellafante's feminism-is-dead critique with the following statement: "We Third Wave feminists should stop having so much fun. We should stop writing books and publishing magazines that people enjoy and that make people laugh out loud. . . . The strategy of producing an embraceable girl culture that allows us both to point out what's wrong with our culture and at the same time celebrate an alternative is clearly just too subtle and complex for the simple-minded grasp of Ms. Bellafante" (qtd. in Brown).

Bellafante and Stoller represent just two voices in a complicated dialogue over who qualifies as a feminist writer and activist. Debates like these, however, are particularly disturbing in their attempt to divide feminist communities reductively along generational lines that also partition writing into categories of work versus play, public versus private, or the political versus the popular. Perhaps these generational tensions signal larger concerns over commodification (for example, the market translation of "grrrl revolution" into "Spice Girl Power"), as well as anxieties over online authorship and intellectual property. Protest over locating feminist agency in the conflicted, fragmented, and at times "trendy" grrrl body is in many ways analogous to concerns over the commercializing effects of Web authorship. Editors of online zine *grrrowl* highlight these concerns in their article "Post-feminism Sell Sell Sell." They observe, "The popular picture of a post-feminist is a stylised nineties chick who is smart, sassy and sexy. Anything and everything goes—especially if it's cyber'd, wired, grrr'd. . . . We're led to believe that post-feminism is about hype and hip whereas feminism is about daggy rhetoric." Here, *grrrowl* editors show how popular versions of postfeminist identity—versions built on erroneous generational differences—are marketable ones, especially in online environments. In order to write effectively and build coalitions in these environments, grrrl editors must therefore balance the politics of the anti-commercial feminist movement and the zine scene with the pro-technology, capitalist ideologies of the Web.

In their research on women and online environments, Gail Hawisher and Patricia Sullivan state that the women in their study often described their e-spaces as "behind-closed-doors, private, all-girls' discussion" (192). This depiction is surprising, they argue, given the highly public and commercial nature of Internet spaces. While Hawisher and Sullivan agree with many grrrl zine editors that "feminists must harness the new technologies to serve their own just political and social goals," they also contend that women who participate in online spaces need to rethink public and private distinctions—that is, how these sites might function as both shelter and battleground, as both resistance and production (195). Whether online or offline, grrrl zine networks are working to confound traditional notions of private and public space in order to articulate an alternative, embodied postfeminist writing subject—a subject mediated by the political, technological, and ideological events of the late 1980s and 1990s as well as by a thriving youth zine counterculture. Seizing the production and distribution practices of countercultural publishing and strategically mixing the popular and political, grrrl zinesters point to the possibility of social change within commercial, academic, and mainstream feminist contexts. Moreover, as critical examples of the limits of academic and mainstream literacy, print and electronic grrrl zines alert us to the important material and symbolic effects of writing as an inter-networked, cultural practice.[7]

NOTES

1. Grrrl zines are receiving more mainstream media attention with recent coverage in *Entertainment Weekly*, the *New York Times* (Fryer), and *Spin* (Powers). St. Martin's has published at least one complete book on the grrrl zine movement (see Green and Taormino) and Penguin Press recently released *The Bust Guide to the New Girl Order*, a collection edited by Karp and Stoller, the creators of the print and electronic zine, *Bust*. The Feminist Press has highlighted the movement with several articles in *Ms.* (Sherman), *Signs* (Rosenberg and Garofalo), and *Feminist Studies*. See also Bernstein and Silberman.

2. In a recent *JAC* article on the contemporary youth crisis, Giroux claims that "educators in a variety of fields, including rhetoric and composition studies, have had little to say about how young people increasingly have become the victims of adult mistreatment, greed, neglect, and domination" (10). He argues that educators must pay more attention to how youth resist and challenge "the complex cultural politics and social spaces that mark their everyday lives" in order to provide them with the necessary intellectual and material resources for active citizenship (10).

3. For a useful discussion of the use of zine writing assignments in composition courses, see Frazier.

4. See Crowley who argues that the development of composition as a discipline has depended on the discourse of student need and its universal writing requirement meant to "socialize students into the discourse of the academy." "Like the narrative of progress," she writes, "the discourse of needs interpellates composition teachers as subjects who implement the regulatory desires of the academy and the culture at large" (233). The diverse desires and abilities of the students themselves thus remain largely ignored.

5. For multiple perspectives on the impact of hypertext writing and theory on composition studies, see Hawisher and Self, *Passions*. See also the special issue of *Computers and Composition* on gender, which offers an excellent discussion of the intersections between feminist theory and hypertext in composition teaching and research (Gerrard).

6. One such example is zinester Summerstein's failure to market her zine to mainstream print publishers. In a recent article, she writes, "The book I intended to write was a modern feminist guide/manifesto, which I planned to call *Generation X: Women Here and Now*.... I wanted to create a platform for young women to share real, honest stories that would inspire and support us and lead us back to our long-forgotten power." Her New York literary agent told her, "Make it fresh! What's Courtney Love talking about? What's Alanis doing? *Just. Make. It. Fresh.*" Another editor, Summerstein writes, "beat the thing into mainstream-ready submission, dividing my ideas into chapters like 'Go for It, But Keep It in Check' and 'Let's Get Real.'" Feeling too compromised, Summerstein finally withdrew her proposal and began writing her book without a contract. Interestingly, Summerstein's story does appear on the Web.

7. This essay has greatly benefitted from a series of generous readings by friends, colleagues, anonymous reviewers, and teachers, including Patricia Harkin, Bill Hart-Davidson, Johndan Johnson-Eilola, Rachael Groner, Lisa Langstraat, Becky McLaughlin, Tom Moriarty, Yasmin Nair, Tim Peeples, Siobhan Somerville, Patricia Sullivan, Jan Wellington, Tom West, Gabriel Wettach, and Mike Zerbe.

WORKS CITED

Albrecht-Samarasinha, Leah Lilith. "Sticks and Stones May Break My Bones." Green and Taormino 3–4.

Bellafante, Ginia. "Feminism: It's All About Me!" *Time* 29 June 1998: 54-62.

Bernstein, Robin, and Seth Clark Silberman. *Generation Q*. Los Angeles: Alyson, 1996.

Bolter, Jay David. *Writing Space: The Computer, Hypertext, and the History of Writing*. Hillsdale: Erlbaum, 1991.

Bourdieu, Pierre. *The Field of Cultural Production: Essays on Art and Literature*. New York: Columbia UP, 1993.

Brodkey, Linda. *Writing Permitted in Designated Areas Only*. Minneapolis: U of Minnesota P, 1996.

Brown, Janelle. "Is *Time* Brain-Dead?" *Salon*. June 1998. http://www.salon.com/media/1998/06/25media.html (9 Nov. 2000).

Burana, Lily. "Grrrls, Grrrls, Grrrls." *Entertainment Weekly* 1 May 1998: 76.

Campbell, Anne. *Girl Delinquents*. New York: St. Martin's, 1981.

Carland, Tammy Rae. "Girl Talk." Green and Taormino 46.

Certeau, Michel de. *The Practice of Everyday Life*. Berkeley: U of California P, 1984.

Chan, Kristy. "Kristy's Riot Grrrl Review." http://www.geocities.com/WestHollywood/9352/human.html (15 Dec. 2000).

——. "Girl Love Is. . . ." 4 Dec. 1999. http://www.geocities.com/SoHo/Cafe/3685/girllove.html (9 Nov. 2000).

ChickClick. http://www.chickclick.com (9 Nov. 2000).

Crowley, Sharon. "Composition's Ethic of Service, the Universal Requirement, and the Discourse of Student Need." *JAC* 15 (1995): 227-39.

Doane, Mary Ann. *Femmes Fatales: Feminism, Film Theory, Psychoanalysis*. New York: Routledge, 1991.

Duncombe, Stephen. *Notes from Underground: Zines and the Politics of Alternative Culture*. London: Verso, 1997.

Emelye. "Cyberscared." Green and Taormino 152-55.

Fraizer, Dan. "Zines in the Composition Classroom." *Teaching English in the Two-Year College* 25 (1998): 16-20.

Fraser, Nancy. "Rethinking the Public Sphere: A Contribution to the Critique of Actually Existing Democracy." *Habermas and the Public Sphere.* 1992. Ed. Craig Calhoun. Cambridge: MIT P, 1997. 109-42.

Fryer, Bronwyn. "Offering Curious Girls Room for Exploration." *New York Times* 21 May 1998, natl. ed.: D9.

Garrison, Ednie Kaeh. "U.S. Feminism—Grrrl Style!: Youth (Sub)cultures and the Technologics of the Third Wave." *Feminist Studies* 26 (2000): 141-70.

Gere, Anne Ruggles. *Intimate Practices: Literacy and Cultural Work in U.S. Women's Clubs, 1880-1920.* Urbana: U of Illinois P, 1997.

——. "Kitchen Tables and Rented Rooms: The Extracurriculum of Composition." *College Composition and Communication* 45 (1994): 75-92.

Gerrard, Lisa, ed. *Computers, Composition and Gender.* Spec. issue of *Computers and Composition* 16 (1999): v-205.

Giroux, Henry A. "Public Pedagogy and the Responsibility of Intellectuals: Youth, Littleton, and the Loss of Innocence." *JAC* 20 (2000): 9-42.

Green, Karen, and Tristan Taormino, eds. *A Girl's Guide to Taking Over the World: Writings from the Girl Zine Revolution.* New York: St. Martin's, 1997.

——. "Zinestresses of the World Unite! Notes on Girls Taking Over the World." Foreword. Green and Taormino xi-xiv.

Hawisher, Gail, and Cynthia L. Selfe, eds. *Passions, Pedagogies, and 21st Century Technologies.* Logan: Utah State UP, 1999.

Hawisher, Gail, and Patricia Sullivan. "Women on the Networks: Searching for E-Spaces of their Own." *Feminism and Composition Studies: In Other Words.* Ed. Susan Jarratt and Lynn Worsham. New York: MLA, 1998. 172-97.

Hanna, Kathleen. "Jigsaw Youth." *Bikini Kill.* 30 Dec. 1999. http://www.columbia.edu/~rli3/music_html/bikini_killjig/saw.html (9 Nov. 2000).

Hanson, Susan, and Geraldine Pratt. *Gender, Work, and Space.* London: Routledge, 1995.

Hellfire. http://www.hellfire.com (9 Nov. 2000).

Hewlett, Jamie, and Alan Martin. "Tank Girl Born in Worthing." *Worthing's Community Web Site*. http://www.worthing.co.uk/tank.htm (9 Nov. 2000).

Horner, Bruce. "Students, Authorship, and the Work of Composition." *College English* 59 (1997): 505-29.

Karp, Michelle, and Debbie Stoller, eds. *The Bust Guide to the New Girl Order*. New York: Penguin, 1999.

Kennedy, Pagan. *'Zine: How I Spent Six Years of My Life in the Underground and Finally . . . Found Myself . . . I Think*. New York: St. Martin's, 1995.

LeCourt, Donna, and Luann Barnes. "Writing Multiplicity: Hypertext and Feminist Textual Politics." *Computers and Composition* 16 (1999): 55-71.

lisa. *bitch/dyke/whore zine*. 22 Feb. 1998. http://members.aol.com/myredself/bdw.html (18 June 1998).

Magnuson, Ann. "Other Voices, Other Wombs." Green and Taormino xv-xviii.

Nguyen, Mimi. "Punk Rock Is Positively Bovine." *Slant E-Zine*. http://members.aol.com/Slantgirl/bovine.html (18 June 1998).

——. "White Liberal (Punk) Feminisms." *Slander*. 31 Aug. 1997. http://worsethanqueer.com/slander/83197.html (9 Nov. 2000).

O., Cindy. "i like things to be small . . . ." Green and Taormino 71-72.

Payne, Michelle. *Bodily Discourses: When Students Write About Abuse and Eating Disorders*. Portsmouth: Boynton, 2000.

Planet Grrl. http://www.propaganda-i.com/planetgrrl/info/contributors.html (9 Nov. 2000).

"Post-feminism Sell Sell Sell." *grrrowl*. 8 Nov. 1999. http://digitarts.va.com.au/grrrowl3/sarah/postt.html (9 Nov. 2000).

Powers, Ann. "Everything and the Girl." *Spin* Nov. 1997: 74-80.

"Riot Grrrls." http://www.geocities.com/Wellesley/6788/main.html (26 Aug. 1999).

Robertson, Pamela. *Guilty Pleasures: Feminist Camp from Mae West to Madonna*. Durham: Duke UP, 1996.

Rosenberg, Jessica, and Gitana Garofalo. "Riot Grrrl: Revolutions from Within." *Signs* 23 (1998): 809-41.

Salkowski, Joe. "Geekgirl Grudge Match." *StarNet Dispatches*. 23 Jan. 1997. http://azstarnet.com/public/dispatches/features/geek.htm (9 Nov. 2000).

Sandata, Sabrina. "A Nice Little Asian Girl." Green and Taormino 98–99.

Sherman, Aliza. "Estronet.com—Next Wave Zine Web Collective." *Ms.* Sept.–Oct. 1998: 41.

Summerstein, Evelyn. "Absolutely Capitalist: Adventures in Mainstream Publishing." *Bitch: Feminist Response to Pop Culture* 3.1. Spring 1998. http://www.bitchmagazine.com/archives/9_99abcap/abcap.htm (9 Nov. 2000).

Vale, V. *Zines!* Vol. 2. San Francisco: V/Search, 1997.

9
CYBER-SPACES OF GRIEF

ONLINE MEMORIALS AND THE COLUMBINE HIGH SCHOOL SHOOTINGS

Maya Socolovsky

> Modern memory is, above all, archival. . . . Fear of a rapid and final disappearance combines with anxiety about the meaning of the present and uncertainty about the future to give even the most humble testimony, the most modest visage, the potential dignity of the memorable . . . who, today, does not feel compelled to record his feelings? . . . the less extraordinary the testimony, the more aptly it seems to illustrate the average mentality.
> —Pierre Nora

> Through the use of today's technology, we have established Perpetual Memorials, a permanent resting place online to memorialize loved ones. Our mission is to be the unsurpassed leader in the public service of celebrating and preserving the memory of those who have passed. . . . Our goal is to maintain these on the Internet throughout ages to come.
> —*perpetualmemorial.com*[1]

In the aftermath of the September 11, 2001 terrorist attacks on New York's World Trade Center and on the Pentagon, the issue of memory and memorials once again surfaced as poignantly and as acutely as it did in past debates surrounding the Vietnam Veterans Memorial, the construction of a Holocaust museum in

Washington D.C., and the Oklahoma City Memorials. With architects, citizens, and city planners more than ever aware of the way in which memory is a product of social history and political agendas, debates over the ruins of the World Trade Center began. While an on-site memorial has taken some time and careful planning to establish, the act of memorializing the losses of September 11 began almost immediately after the events. For example, less than two weeks after the attacks, T-shirts, models of the Manhattan skyline, and American flags sprang up around the country. Commemorative books and magazines also began to appear, all capturing, dissecting, and agonizing over the moment of the attack.[2] If, as Pierre Nora argues above, monuments are built in place of memory, allowing us to displace the location of memory so that we do not have to hold it within ourselves, the nation's attempts to deal with grief by immediately displacing it onto patriotic consumer objects of memory signal several things. The immediate desire to monumentalize suggests an anxiety about and inability to process grief. It demonstrates a strong resurgence of community and a desire to translate one's private personal voice into a collective voice. It represents the need to retrieve unspeakable absences and create presences in their place.[3] Finally, it suggests an impatience with, and fear of, the intangibility of loss, the otherness and absence of death, and the sheer incomprehensibility of endings. The speed and immediacy with which an event comes to be memorialized is part of today's memorializing strategy. Although the desire to de-other death and to translate its absence into presence is an inevitable feature of most traditional and even contemporary monuments and memorials, I argue here that this desire is a particularly marked and nuanced feature of Internet memorials. As a result of this attempt to hide the gap between self and other, and between the bereaved and the dead, the Internet memorial lacks an othered and haunted space in which ghosts, as the most affective carriers or signifiers of memory, might reside.

In recent years, the Internet has become one of the sites of memory, that, like a museum or monument, Nora would lament signals a loss of "real" or natural memory in modern times.[4] Even aside from the web memorials that have begun to appear in cyberspace, computers have long taken on the tasks of memory. Edward Casey argues that "we have turned over responsibility for remembering to the cult of the computers, which serve as our modern mnemonic idols. . . . Human memory has become self-externalized: projected outside the remembered himself or herself" (2). With the undeniably significant growth of Internet use among middle-class sections of the population, and the use of the Internet as a site of memory, we have to ask ourselves how this technological media affects the structure of memory and the ways we understand and experience our temporality and our mortality. Certain features of the Internet and of computer usage affect the way memory is written online. Because words and images are digital rather than physical, the electronic text is always a simulacrum with no concrete tangible instantiation of itself elsewhere except as mediated through a representing device. Any webpage invites its viewers or readers to sample, rather than master, the online text, and the cursor, signaling the reader's presence, can change the text itself, processing and manipulating what appears and replacing one page with another, so that no final or fixed version of an online text actually exists. As such, no Internet experience has a center or a periphery, and although we may psychologize homepages as the centers or beginnings of our online experience, a crucial part of being online is the constant vacillation of a page's structure, words, or images. Furthermore, it is our interaction with the text that brings about such oscillation.

Thus, as a highly interactive, continually replaceable, reader-oriented medium, we have to ask what kind of consoling space the Internet can provide for our collective and private memories. Internet readers or surfers are frequently understood as being themselves dispersed and multiple:

> In the mode of information the subject is no longer located in a point in absolute time/space, enjoying a physical, fixed vantage point from which rationally to calculate its options. Instead it is multiplied by databases, dispersed by computer messaging and conferencing . . . dissolved and materialized continuously in the electronic transmission of symbols. In the context of Deleuze and Guattari, we are being changed from "arborial" beings, rooted in time and space, to "rhizomic" nomads who daily wander at will . . . across the globe, and even beyond it through communications satellites, without necessarily moving our bodies at all. (Poster 15)

If the body no longer serves as a limit to our subject position or experiences, then sociable interactions and acts of empathy online are mediated through a faceless and bodyless community. Citizens of the Internet do adapt to the virtuality of their space: "As participants adjust to the prevailing conditions of anonymity and to the potentially disconcerting experience of being reduced to a detached voice floating in an amorphous electronic void, they become adept as well at reconstituting the faceless world around them into bodies, histories, lives" (Porter xii). That is, a virtual space becomes a place of belonging and even of collectivity. Web memorials in particular are especially anxious to interpellate themselves as archival places (rather than virtual spaces), "resting places" for the dead that can be visited with ease by the bodyless and faceless bereaved. The Internet thus becomes a gravesite, a place of departure and presence that is interactive and personal enough so that those in mourning can manipulate the details of the memorial and at the same time experience it as a meeting-place for lost loved ones to communicate with them.[5]

It can of course be argued that physical memorials also act as sites of departure and mourning that give a presence to death, and that, perhaps unlike online memorials, they are more affected by the public's need to rationalize and maintain power. In physical memorials, loss and absence take root in the concrete and the physical so that their intangibility can take on a shape and form that is known, sensed, and understood. In such ways, we stave off the possible nothingness and unknowingness of mortality. We live with a shared "museal consciousness" that "understands the significance of collecting, ordering, representing, and preserving information in the way that museums do" (Crane 2). As James Young notes, certain varying narratives (political, national, aesthetic) emerge when we look at the diversity of sites of memory. While traditional physical monuments are usually read as relieving us of the burden of remembering, counter-monuments, in which the monument in various ways foregrounds its own impossibility of containing memory, defy our impulse to write presence into absence and to domesticate death through narrative.[6] That is, some physical memorials highlight the unrepresentability of death and the amnesia of memorialization, and some do not.[7] Crucially, unlike online memorials, all physical memorials have spatiality, and present a sense of lived space that retains a distance between the observer/visitor and the memorial. This distance serves as an enactment or reminder of what lies outside signification; that is, it indicates the silence and absence of death. When we visit a physical memorial, the boundaries of our body always serve as reminders of difference and otherness. Even in a highly interactive memorial such as the United States Holocaust Memorial Museum, where visitors are architecturally removed from an American space and encouraged to enter an alternate experience and space, the gap between self and other (the past, death, nothingness) is retained. Physical memorials or museums realize the limits of representing absence, and in lending a physical dimension to remembrance, they are always asking themselves, "What kind of architecture could do justice to an event that resists profound aesthetic expression?" (Linenthal, "Locating" 220–21).

The political debates surrounding the design of Maya Lin's Vietnam Veterans Memorial, and the public responses to the Wall, also show a contradiction of narratives that reveal both people's desires to make death dialectical, to bring back the dead, to interact with the dead, and also, to figure death as unknowable and absent. After the dedication of the Memorial on November 11, 1982, critics of the design described it as grave-like, as emasculating and unpatriotic, and as something that induced grief and sorrow rather than redemption (Wagner-Pacifici and Schwartz). In that sense, the Memorial was read as failing to narrate a heroic story of war, and as seeming instead to go directly into a space of memory and absence that some of the public was reluctant to enter. At the same time, objects left by visitors at the Wall–such as medals, bottles, photographs, dog tags, and letters–can be read as acts of self-expression and self-extension, as people interact with the memorial and self-consciously write themselves into a national memory and history. This archival impulse reflects a desire to make oneself immortal, and to cross the boundary created by time, place, and experience. Internet memorials, however, see themselves as always crossing time and place, and thus, implicitly, as also traversing the boundaries of experience so that the boundaries or gaps are not articulated, and, to an extent, cease to exist. The Oklahoma City Memorial includes The Field of Empty Chairs, where 168 chairs are placed empty as reminders of life lost, articulating the absence that friends and family feel, but also in themselves creating a space of haunting and ghostly absences (Linenthal, "Memory"). Physical memorials awaken ghosts, by introducing the viewer or visitor into a space of ambivalent absence, gaps, and difference. Whether the memorial tries to narrow or hide that gap by narrativizing death and crossing borders of experience, or whether it consciously emphasizes the gap and displaces death and absence as ultimately unrepresentable, the physicality of the space occupied by monument and observer will, to some extent, always highlight the unknowability of absence and the otherness of death in a way that Internet memorials cannot.[8]

Sometimes writing themselves consciously against the potential decay of physical memorials, web memorials describe themselves as the most recent, and the most effective, form of memory for our time. To an extent, web memorials can and should be read as museums. Susan Crane writes, "To each era its own forms of memory: the recent and explosive evolution of the Internet, like a museum, like any of the prosthetic cultural devices created to supplement mental memory functions, offers an externalized technologized memory" (12). In talking about web memorials, museal discourse must be redefined:

> A "museum" may be any real or imaginary site where the conflict *or* interaction *or* simulation of or between personal or collective memory occurs. Museums are more than cultural institutions and showplaces of accumulated objects: they are the sites of interaction between personal and collective identities, between memory and history, between information and knowledge production. (12)

Hence, web memorials are not only sites of unhaunted memory, but also sources of information for grief counseling, religious education and inspiration, community-building, and various clearly laid-out political agendas and voices.

The homepage of virtualmemorials.com introduces itself as follows: "We create memorials that celebrate the lives and personalities of those we have lost and provide a place where these cherished images and biographies will have a permanent home." The "we" here collapses the distinction between public and private: a family's personal loss can become, through the memorial, part of a national loss and thus subject to national grief. Like a museum artifact or object, the images and photographs of the deceased, who are ordinary members of the public, affirm the significance of the life lost. As Crane writes, "Being collected means being valued and remembered institutionally; being displayed means being incorporated into the extra-institutional memory of the museum visitors" (2). The reader is then invited to visit a memorial by clicking on any of the many names that appear on the rest of the page, or on one of the three photographs pictured beneath the caption. Once inside a memorial page, you can choose to enlarge the photograph, or to read the biography, travels, reflections, or passages about family life. You can sign the Guest Book, learn about support groups, click on affiliated pages, and read more about "us," the memorial website.

Crucially, the homepage sees the web as providing a permanent place for memory, even though it also uses the changeability and easy replaceability of images on the web to its advantage. For example, after September 2001, virtualmemorials.com added a new sidebar (now removed) that expressed sympathy to those affected by the events of September 11 and a hope for peace and an end to violence. The site thus wrote itself into a national dialectic of mourning and grief. At the same time, a single click would eliminate this message of sympathy and take the viewer to an affiliated page such as barnesandnoble.com in order to purchase books such as *The Day Diana Died* and *How to Survive the Loss of a Love*. Because all pages on the web memorial are the same distance away from the reader—one click of the mouse—and are all equally vulnerable to the viewer's decision simply to exit the site, the idea of an archive of memory has to be redefined. Archival memory on the Internet comes to mean collection and display, but above all, replacement. Although loss or absence are ostensibly at play when we make one image disappear without a trace, the fact that the image is always instantly replaced with another paradoxically means that our experience of loss and death is one of presence, even of an excess of presence. Because there is no distance between the different presences on the screen, and no temporal narrative between pages, the memorial, even as it speaks of death and shows photographs of the dead, lacks any trace or haunting.

A memorial archive on the web also overtly articulates the permanence of memory through this medium, and reiterates its presence as a storehouse for archives and artifacts. Virtualmemorials.com repeatedly assures us that its archive and memories are permanent. It is "a place for reflection and enduring memories to be passed on generation after generation" and also "a place where future generations can learn about their ancestors long after original records, photographs, and writings have been destroyed." The creator of virtualmemorials.com realizes that "Our lives extend beyond mere numbers on a tombstone" and says that "the Internet and its new technologies have given us the ability to preserve in colorful detail the chosen highlights of our lives as we have never been able to before." This form of memory offers real documentation, it seems, and an attention to personal detail that turns the anonymous victim of a car accident into someone we feel we know. The site chooses not to linger on how the very newness of the web might make it vulnerable as a medium for concrete memorialization. At the same time, as explorers of an online museum, we cannot help but be aware of the temporary nature of the memorial and of the ways the medium tackles its own temporality. For example, on December 18, 2001, in the Guest Book, the following message appeared: "Apologies for the system being down for two weeks. Virtualmemorials was hacked and its servers were taken offline."[9] Loss of memory online, however, still does not entail actual absence. Rather, unlike the defaced ruins of physical markers of memory in cemeteries or on tombstones, a recently kaput online memorial is still full of presence. Just after September 11, for example, condolences.com, showed only an American Red Cross sign, with the motto "Together we can save a life," and an American flag. Beneath that appeared the caption "this domain is for sale. please make an offer."[10] But even this website was not entirely static: clicking on either the Red Cross sign or the flag would bring you to amazon.com and its statement about the Red Cross Disaster Relief fund which asked for and enabled you to make a donation. In July, 2003, the site simply announced, in white letters against a black background: "condolences.com FOR SALE make an offer." No ghosts of the past emanate from the remains of this memorial website. Instead, it can only be a comment on events of the recent past: the earlier request for a charitable donation reminded us of recent national tragedy, and the request to buy the domain emphasizes the fragility of real estate on the Internet.

One's experience of the web memorial undoubtedly depends on whether the reader or viewer knew the loved one memorialized online. It is possible to stumble unintentionally onto the pages of a web memorial, just as it is possible to accidentally leave them behind quite suddenly. A web surfer who finds him or herself wading through others' publicly personal expressions of grief is encouraged, as a visitor, to leave personal recollections and tributes and to interact with the memorial website. He or she becomes a voyeur of bereavement, of loss, and of death, and in this way, the experience of visiting an online memorial without already having a connection to the departed becomes emblematic of our desire to have or know death collectively. Much as cars slow down by the scene of an accident and stare, visitors can gaze at the privacy and intimacy of this medium of public grief; they can experience a safe but curious proximity to death without undergoing it themselves. While driving past the scene of a car accident may existentially remind one of the unspeakable horror of death, the narrative of life and death online serves rather to tame that horror and to give it a reassuring voice and presence that is palatable, understood, and known.

For those who have placed a memorial online and return to read it and dwell with it, cyberspace, as a resting "place" for the departed, serves as a point of contact between the world we know and the afterlife. One memorial poem reads: "It will soon be four years since you went away and it still seems like yesterday ... until we meet again, precious one" (virtualmemorials.com). Another memorial in the same site talks of the remembered one as an angel. The Internet space becomes a heavenly sphere and a waiting place for the departed "angels" to await their loved ones who are still on earth. On December 26, 2001, one visitor wrote, "I don't know what I would have done if I hadn't found this site. I know that I can come in and it is like I can talk to my mom and I feel so much better after I do talk to her."[11] Others directly address the dead: "we miss you. . . . I look so forward to seeing you again."[12] Web memorials write the dead into cyberspace, but these dead are not "other" to the living. The space they occupy is not "other": it is a place that the bereaved can simulate being in, can signify and refer to, and can know. An article from ABCNews.Com about web memorials that appears on virtualmemorials.com reads, "Perhaps it was inevitable that just as more and more people seem to live online, others are dying there too. For the dead, cyberspace cemeteries serve as memorials; for the living, they are a place to mourn, to offer condolences, to recall other losses." Being dead online emphasizes death as a narrative, and suppresses its fearful otherness: the lack of spatial distance means we can feel closer to the dead. The lack of physical barriers—stone, concrete, tombstones—means that the gap between the living and dead vanishes and there is no space for emptiness. Instead, death becomes a waiting, and an embrace, figured through language and proximity. For the memorials that offer religious faith as a condolence (and the majority of them do), God, heaven, the departed and the living are all cradled—bodiless—in cyberspace.

The bodilessness of cyberspace, for both the departed and the bereaved, has been cited by most web critics as part of the boundless freedom of Internet existence. In calling this freedom into question, Slavoj Žižek asks, "Is not the notion of cyberspace a key symptom of our socioideological constellation? Does it not involve the promise of false opening (the spiritualist prospect of casting off our 'ordinary' bodies; turning into a virtual entity which travels from one virtual space to another)?" (130). For Žižek, this departure from the body and subsequent spiritualist living is actually part of a myth of cyberspace. On the Internet, the "Real"—that which lies beyond language, experience, and the symbolic order, and therefore cannot be expressed—is the digital universe, a series of virtual and binary bytes. It lacks otherness because of its virtualization. Although we Žižek may fear that being online will diminish our contact with real bodily others, the real problem, Žižek suggests, is that the virtualization of the Internet "cancels the distance between a neighbor and a distant foreigner" because "it suspends the presence of the Other in the massive weight of the Real" (154). The Otherness—whether it is the otherness of death, foreignness, absence, or God—is eliminated, in effect, by the collapse of spaces and differences. What we lose is the unrepresentability (otherness) of the Real.

In wandering through online memorials and being voyeurs of death, we are encouraged to think that we can imagine our death or someone else's death. But the death we get online lacks the very thing that makes it unrepresentable and thus unknowable. While a physical memorial contains gaps, an Internet memorial says it all. Death is narrated fully, and although the departed are mourned and missed, death itself is understood and mastered. Thus, when the voids are filled in on web memorials, a different kind of loss occurs because the elusiveness of death that usually resides in absences has been articulated. The virtual reality of the Internet fills in the absences to such an extent that ghosts vanish, and we are left with an "excessive fullness." As Žižek says,

> The commonplace according to which the problem with cyberspace is that reality is virtualized, so that instead of flesh-and-blood presence. . . . We get digitalized spectral apparition, *misses the point*: what brings about the "loss of reality" in cyberspace is not its emptiness (the fact that it is lacking with respect to the fullness of the real presence) but, on the contrary, its very excessive fullness. (155)

The problem, therefore, is not that cyberspace lacks bodies and involves only an encounter with digital phantoms. Rather, "cyberspace is *not spectral enough*" (155).

What occurs through digitalization is "the almost perfect materialization of the big Other [death, memory] out there in the machine" (Žižek 164). Andreas Huyssen reiterates this with specific attention to our culture's attitude toward memorialization. He writes that "the obsessive self-memorialization per camcorder, memoir, writing, and confessional literature . . . can be said to function as key paradigms in contemporary postmodern culture" and that "far from suffering from amnesia, we suffer from an overload of memories" (253). Our experience of temporality has altered: material life has quickened and media images and technology have speeded up. As Huyssen notes, "speed destroys space and it erases temporal distance" (253). The speed with which pages are eliminated and replaced on a web memorial eliminates space, and the ambiguity and otherness of death. Because we can call up the past on the screen at any time, historical continuity or discontinuity give way to "simultaneity of all times and spaces readily accessible in the present." Consequently, "the perception of distance, both spatial and temporal, is being erased" (253–54).

In this way, web memorials offer us sites of death that are excessively full, articulate, and understood; and while it is in one sense comforting to finally know death through the safe haven of a computer monitor, the death we know does not trouble or disturb in the same way, because in its sanitized form, it lacks the crucial horror of a void. For example, because the creator of virtualmemorials.com feels that "our lives are so much more than the little dash between two numbers of a tombstone," the void or gap signified by that dash is filled in and materializes into, to name but a few, biographies, photographs, recollections, and poems.

Nora writes that our modern memory "relies entirely on the materiality of the trace, the immediacy of the recording, the visibility of the image," and that this concretization of absence and loss is the result of an obsessive anxiety about disappearance (290). Online memorials counter this fear of disappearance by making even the most ordinary citizen a celebrity. If a life is important enough to be documented publicly, then that life had meaning and was not wasted. Online, anyone can be a celebrity. Virtualmemorials.com's text-only memorials are free, and after that prices range from $35 to $225, depending on whether you choose the basic, classic, deluxe, or premium memorial. Each memorial is allowed up to a certain number of photographs, words of text, reflections pages, email links, and a limited number of free updates. The custom-made memorial, whose price depends on each personal design, allows unlimited photographs and texts as well as other multimedia (music, video). The web memorial is, according to Ben Delaney, president of Sausalito, California-based CyberEdge Information Services, a way "for them to say, 'Hey, I'm here, I was here, I made a difference, . . . a way to show others that these people existed.'"[13] The technology, according to the website, offers something primal: a path to immortality: "everybody hopes for some level of immortality and everyone craves their 15 minutes of fame . . . this is their way to get it."[14] At the bottom of each page, we can read how many times the page has been viewed, and thus measure the extent of the deceased's immortality and celebrity. The different memorials within the site do not vary a great deal: ironically, all are similarly concerned to convey a lost life that was both unique and universal, and to forge a memory of a loved one that suggests being exceptional yet also ordinary. Some of the memorials are heart-rending, showing children and young teenagers lost through illness or car accidents or suicide, and the similar sentiments and layouts of the memorials suggest that a virtual community of bereaved parents, friends, and family is maintained through the website. The speed and immediacy of the web memorials is double-sided: one person's memorial can, at a click, be made to disappear and at the same time be replaced by another person's memorial, putting the two in tandem with one another.

Part of the web memorials' apparent confidence in their own permanence and ability to signify death and loss comes from the religious faith they demonstrate online. Many of the biographies note that the deceased was baptized or is, for example, now "cradled in the arms of Christ."[15] The memorials are supposed to be inspiring, and work as an emotionally healing outlet for grief, but they also serve to make God imminent. Just as death loses its otherness, so does God—or, in most cases, Christ—lose any fearful transcendence or distance. Some pages are almost evangelical in their effort to help others. For example, part of a memorial for a suicide teenager includes a biography that tries to offer help to other parents or teens. It asks the reader to call the number on the website, and "most importantly, place your trust in Jesus Christ." It continues, "we were never designed to tackle this world by ourselves—God is simply waiting for the invitation into your life to walk beside you. From the moment he created you, God has had a special purpose and an awesome life planned for you."[16] God, like death, is comfortably and easily signified and known.

The web memorials developed by parents of the children killed in the Columbine High School Shootings of April 20, 1999, in Littleton, Colorado, share a great deal with the memorials of sites such as virtualmemorials.com. In particular, the way in which Christianity often forms the foundation for memory, and the way in which the websites also serve as informational links and sources, bring the memorials of cassiebernall.com, danielmauser.com, and racheljoyscott.com into the realm of ordinary web memorials. However, from the outset the Columbine memorials are different in terms of celebrity status. While people remembered in virtualmemorials.com depend solely on the website for their celebrity, the work of remembering the victims of Columbine High began before their parents memorialized them online. The news media and national grief and concern that emerged after the shootings all brought the thirteen victims into America's homes, giving a face to death and loss as pictures of the dead were displayed on the front covers of national news media. Thus, the implications of the Columbine web memorials differ. Often using sources outside the Internet as a departure point for memory, these web memorials are part of a national and public narrative that began outside the Internet. The public's scrutiny of and sometimes morbid fascination with the victims means that many of those who visit these websites did not personally know the victims but come to the site intentionally, whether for research into the shootings in general, into the gun control and violence in school debates that emerged out of the killings, or for religious inspiration.[17] In this way, the Columbine websites are a site of public grieving and serve as monuments to what in a sense was regarded at the time, and self-consciously, as a national loss symptomatic of the entire nation's troubled youth. The questions to be asked of the Columbine web memorials, therefore, are: How do they figure and narrate national loss and death? How do they write themselves into the nation's rhetoric of memorials? And, finally, how do they situate themselves as sites that have a responsibility to inform the public, mourn with it, and position themselves as representative voices?

A few days after the Columbine shootings, a community memorial service took place in Clement Park in Littleton, Colorado. Although the service was supposed to include representatives of various religious communities in the area, many felt that the overall tone was evangelical and exclusive of denominations and faiths that were not fundamentalist Christian. Even in the aftermath of the tragedy, as Stream points out, dissent surfaced as critics of the service were accused of "politicizing" the memorial service. That is, from the point of view of the Christian right, an opposition was established: the religious Christian angle represented a "true" and "unbiased" memory; the secular angle by default "politicized" memory. This opposition is crucial in the narratives of the Columbine web memorials, as well as in most of the debates that surfaced after the tragedy. The Jefferson County School District, for example, invited members of the Littleton community to make memorial tiles to decorate the hallways of the school as a permanent memorial. Although initially the tiles were not supposed to include any religious symbols, this restriction was eventually lifted. After the tiles were put up, many of them with Christian symbols, authorities changed their mind and removed the religious tiles. The debate—one's right to freedom of expression versus the separation of church and state—and many of the parents' desires to demonstrate their right to express religious beliefs, could be resolved online, where private memorials could depend on religious faith and at the same time, because of Columbine's status as a national tragedy, serve as a public monument for those killed.

The homepage and entire memorial of Cassie Bernall, at cassiebernall.com, is figured around the moment of Cassie's death and her response to the idea of death. Allegedly, when the killers pointed their guns at Cassie they asked her if she believed in God, and she said, "yes." Her mother, Misty, wrote a best-selling book a few weeks after the tragedy, entitled *She Said Yes: The Unlikely Martyrdom of Cassie Bernall*, which is heavily promoted throughout the memorial. The inspirational book, which has received awards and gained widespread recognition, turned Cassie into a teen-idol and reportedly helps teens and their parents work through difficulties.[18] At the top of the page is a graphic that says "Cassie Bernall She Said Yes," with the "Yes" enlarged as it is on the cover of the book. The killings at Columbine High School are introduced immediately underneath: "On April 20, 1999, Cassie Bernall, a junior at Columbine High School in Littleton, Colorado, was a typical teenager having a typical day; then a classmate trained a gun on her and asked if she believed in God. She said 'Yes.'" Cassie's memorial is centered on her "Yes," marking her death as a beginning and an affirmative presence, rather than an inarticulate void. On the right-hand side of the page is an image of the book, with toll-free phone numbers to enable the visitor to buy the book and get free shipping, to order a video of the book, or to buy it at either Amazon, Barnes and Noble, or Borders. The rest of the homepage features side-bars that can be clicked on to find out about "The Story of the Best-Selling Book," to learn more about the book (we can read the first chapter and reviews) to read responses to questions about Cassie and to access links to other Columbine victims. In December 2001, in response to the controversy that surfaced around the story of Cassie's "Yes," the viewer could also click on the question, "Did Cassie Say Yes?" and find out by linking to an article by Wendy Murray Zoba of *Christianity Today*.[19]

Because Cassie's memorial depends on her affirmative signification of death—"yes"—it concerns itself with the possibility that the story is not true.

However, without the "yes" at the crux of this particular memory, the memorial's redemptive strategy risks disintegration. The "yes" assures Cassie's parents:

> We know that God is working good out of this horrible nightmare. . . . We know that our daughter was no saint. She was far from perfect, but she was prepared to die for her faith in God. Her final word "yes" will always be a challenge and inspiration to us. Our hope is that her "yes" inspires others to take their faith more seriously.[20]

Cassie's affirmative response gives her a voice after her own death that does not even have to wonder whether her life was wasted and meaningless. In terms of the memorial, her life is not over just because of her death; rather, the moment of her "yes" brings her closer to God. The web memorial thus serves to demonstrate the shared place that Cassie, God, and her parents still occupy, and the de-Othered nature of her death. Misty Bernall, in a link that takes the viewer from the web memorial to an article entitled "Voices of Columbine: The Family of Cassie Bernall," reports that some of the kids who were in the library at the time of Cassie's murder are "one-hundred percent sure" and responds to others' doubts by saying "some people are very cynical" (Wallace). Tackling the same issue, Zoba's piece in *Christianity Today* does point out that testimonials depend on the witness's ability to process events at the moment of trauma, but in the end prefers to wonder why people are so interested in debunking Cassie's account and affirmative "yes."

What emerges from a reading of cassiebernall.com is that it is crucially important for the memorial to be able to give a presence to the moment of Cassie's death, and in this way to know and understand it. While her "yes" signals her faith, the public and private focus on her "yes" signals the need to narrow the gap between life and death, to be present at the moment of departure, and to make it manageable and narratable. The story, circulated continually, allows us to gaze on the moment of death and even imagine brushing up against it without having to come too close. Significantly, her death is remembered as occurring not in a void, but in a place filled with presences: courage, faith, and God, and like the more religious memorials in virtualmemorials.com, cyberspace is interpellated as a place where, through death, an encounter with a knowable intimate God is possible.

Like cassiebernall.com, Rachel Scott's web memorial also centers itself on a religious message that was prohibited from appearing in the hallways of Columbine High School. An earlier version of the memorial webpage (December 2001) showed a photograph of Rachel at the center of the homepage: the same photograph of her that appeared in newsmagazines and in the media after the killings, but enlarged and more detailed. In the memorial she sat, smiling, with her head tilted and body leaning to one side in what is almost a sensuous pose, and the background behind her consisted of enlarged handwriting, of which the viewer could just about read the first few words: "Dear God."[21] Two columns ran down the side of the photograph with options for further exploration in the memorial. Under the photograph there was a quotation from Rachel: "Don't let your character get camouflaged with your environment. Find who you are and let it stay in its true colors." At the time of this memorial, the site was undergoing development; a caption on the homepage read: "A new site for Rachel is still currently under development," but the viewer could still click on links that allowed them to contact her family, make donations, see some of Rachel's art, and learn more about her.[22]

Unlike Cassie's memorial, this one, in the immediate aftermath of September 11, also invited the viewer to click on a heading that read, "In response to the Tragedy" of September 11. The new page that then appeared expressed condolences and urged the viewer to turn to God and unite in prayer: "Let us not forget how much bigger God is than what happened on that fateful day. . . . With our faith in Jesus Christ, we will rise victorious and become a stronger, more loving nation that sets an example for the world just as Rachel did." Rachel was thus written into recent national events: her death and the deaths on September 11 were brought together to create, implicitly, a sinister connection between the events. The teen terrorists, Klebold and Harris, according to Harris's diary, fantasized about flying a plane into a New York skyscraper.[23] Rachel *is* the nation, and in her memorial death and absence are again turned into presences: determination, faith, and redemption. On today's website, Rachel continues to be rhetorically aligned with the nation through a link to a site entitled "Pray for President Bush." Through this link, one reaches a homepage full of images of George W. Bush, and flags flying, that describes itself as a "non-denominational non-partisan ministry dedicated to lifting up President George W. Bush in prayer as he serves this country as President of the United States." The site sanctifies Bush and his relationship to the U.S. by quoting from Romans 13:1 ("Everyone must submit to the governing authorities, for there is no authority except from God, and those that exist are instituted by God"), and notes that it is God who "has placed" Bush "in the office of the Presidency." The site is interactive, asking viewers to let it (the website) know that they are praying for Bush. The fact that this entire page is linked to Rachel Scott's web memorial suggests that the writing of her memory continually involves the creation of a rhetorical relationship between her and God, her country, and her faith.[24]

Of the three Columbine memorials, racheljoyscott.com is, in terms of its graphics and aesthetic layout, the most similar to a traditional memorial plaque. The biography lists her favorite things in order for us to know her, as do other web memorials, but it does so in the form of asides, while the central text—a contemplative biography that focuses on her devout Christianity—describes her personal relationship with Jesus Christ and the faith and love that she wanted to share and witness to others. The biography also describes her as "a girl from Littleton, Colorado, who said 'yes' to God, everyday, even in the face of death," implicitly referring to and echoing Cassie Bernall's own martyrdom but also repositioning Bernall's faith as one that pivots around a singular word and event (her "yes" at the moment of death).[25] In turn, this frames Scott's own faith as apparently more authentic, less commercialized, and more pervasive than

Bernall's. The tension between Scott's desire to witness, and her quotation featured on the earlier homepage—be yourself and stay with that; don't be influenced by your environment—is not addressed in any of her memorials. Instead, the memorial pages function as inspirational sources, and as messages of hope. Death and tragedy, they suggest, are beginnings, not endings. Like Cassie's memorial, Rachel's also refers to a memoir that lies outside her web memorial, but promotes it more lightly. *The Journals of Rachel Scott: A Journal of Faith at Columbine High* (by Beth Nimmo—Rachel's mother—and Debra Klingsporn) appears under a link to "Products," which in itself links to amazon.com's page, and to another book written by Rachel's father: *Rachel's Tears: The Spiritual Journey of Columbine Martyr Rachel Scott*. This latter book appeared on the homepage of the earlier Scott web memorial, featuring excerpts from her private journals, and asking, in its byline, "Was there a prophesy in Rachel's Tears?" It added, "Her life ended! Her legacy began!" Like Cassie's memorial, Rachel's various memorial pages center on presences: a certainty of faith and proximity to God, and on the fateful inevitability in her death, which lends more credence to the idea of God's personal role in bringing it about.

While cassiebernall.com and racheljoyscott.com function as memorials that narrate death as an affirmative presence and see it as their responsibility to educate and inform the public through evangelical rhetoric, Daniel Mauser's web memorial stands out in honoring the deceased through a directly politicized form of memory. Like the others, it is clearly dedicated to him as a victim of the shootings, and features links to his life, a photo album, memories, guest book, and words of comfort. But unlike the others, this memorial features links to debates about guns and violence and to photographs of the other Columbine victims.[26] The memorial attempts to educate and inform the viewer about the event at Columbine as a whole, and not just on the way it struck this particular family. In this sense, it speaks in national terms and expresses national anxieties, but without an explicitly religious overtone. At the bottom of each page of this memorial are the words "we are all Columbine!" embedded between the image of a memorial ribbon and three columbine flowers. As a signature to whatever has come before it, this emblem serves to unify all the various aspects of the memorial (which makes wide use of the Internet's hypertext abilities) no matter whether it appears beneath photographs of Daniel and his sister as children or articles about the gun control debate. This bringing together of the personal memorial with the public domain collapses the distinctions between the two, articulating the grief over Daniel's death also as grief over the wide dissemination of guns in the U.S. Its message also draws in the viewer, nationalizing the specifics of the tragedy, while at the same time keeping them intimate and domestic with the graphics. That is, through this memorial, all of America suffers from the Columbine syndrome and *is* the violence and tragedy of the event, but the universal sharing of responsibility also implies a communal bond and collective survival in the face of teen anger.

Through the memorial, we learn about the HOPE Columbine Atrium and Library Fund—the project designed by parents of the Columbine victims to tear out the library and build a new one—and Tom Mauser's work with SAFE Colorado, an organization that pushed for Amendment 22, the law that closed the gun show loophole in Colorado. The links to pages that discuss gun control are extensive, featuring among other things letters written to U.S. Senators regarding gun issues, and Tom Mauser's own discussion of and responses to gun clichés. It becomes clear that the political nature of this memorial has elicited negative responses from the NRA and from other gun defenders, so much so that, according to information on the memorial, Tom is featured on "Wanted" posters for the NRA and receives hate mail from them. The guest book, in the past, has been home not only to expressions of sympathy toward the family, but to slurs, insults, and attacks from pro-gun individuals; and the website has also featured a response to these attacks from Tom Mauser.[27] For example, under the heading "Why the Personal Attacks? Why the Hate Mail?" Tom discusses visitors' messages that attack him, and he quotes some of them. He notes that in writing to these people, "some said they didn't mind my publishing their names and words herein—they just see it as a 'badge of honor' in promoting their beliefs." The memorial website thus becomes a voice for the opposition, as Terry Chelius of Whitewater, Colorado writes, "Get a life. . . . This is a great vehicle to get your fifteen minutes of fame, but try to get on with your life . . . get a job and buy a good gun. . . . Tommy boy, you and Linda have ridden the tragedy like a roller coaster, never missing an opportunity to get your face on TV or in the papers." Or Richard W. Pope of Des Moines, Iowa, who calls Tom "a weak pitiful man . . . who is trading on the dead body of your son . . . to get your name in the paper." Another accuses the Mausers of "using the death of your son to desecrate the constitution . . . Quit using your son as a political tool." The implication, clearly, is that memory has become politicized because it no longer expresses or reiterates a "truth" or fundamental preconceived right or belief. That the Mausers have chosen to leave the slurs there and make them part of the memorial suggests that a conscious part of the memory-work in this memorial is the understanding that no memorial is apolitical.

Daniel Mauser's extensive memorial also features descriptions of his vigil, funeral, and burial procession. In referring to these outward events of death and bereavement, the web memorial refuses to offer just one narrative of death. But we do learn that Tom and his wife barely visit the grave "because of painful memories." The absolute absence of the physical gravesite is countered by the presences that fill his website memorial, where activism, information, and education, as well as a careful expression of personal religious faith, and narrative biographies and memories, help to mediate and temper the void and silence of death. All the Columbine memorials thus nationalize their private grief and loss by writing in collective and informative voices. Whether their agendas espouse religion or gun control, they reenact the fundamental "excessive presences" of all web memorials: eliminating undialectical death and filling the space between the bereaved and the deceased with the politics of protest and prayer.

NOTES

1. www.perpetualmemorials.com. December 2001. As of July 2003, this website no longer exists.

2. For example, some T-shirts showed the Manhattan skyline and read "The may destroy our buildings, but they will never destroy our faith—in memory of lives lost on September 11."

3. Don Handelman and Lea Shamgar-Handelman write that "death necessarily turns presence into absence," and they argue that the dead are used for the needs of the living (3, 4).

4. Nora explains "real" memory as follows: "We have seen the end of societies that had long assured the transmission and conservation of collectively remembered values, whether through churches or schools, the family or the state . . . 'real' memory—social and unviolated, exemplified in but also retained as the secret of so-called primitive or archaic societies" (284–85).

5. The continual replaceability and instability of the web are one of the challenges I faced when writing about web memorials. Far from being set in stone or being in any way permanent, pages are updated, new information is added, and sometimes old information vanishes. My paper is based on removing the website from its location online: I printed out the pages and studied them as if they were paper-based texts. Thus, in writing on it, I had to make it, to an extent, "unInternet," freezing the information of the moment.

6. An example of a countermonument is Harburg, Germany's *Monument Against Fascism*, designed by Jochen Gerz and Esther Shalev-Gerz. It was unveiled in 1986, and consisted of a black pillar that over time was sunk into the ground. Visitors were invited to inscribe their names on the pillar. The monument thus demanded interaction from viewers, and challenged traditional monuments by suggesting, through the lowering, that memory consists of an absent monument rather than a present one. See Young 29.

7. Katie Trumpener points out that "writers have become . . . critical of older public monuments and ceremonies, as chilly abstractions from lived experience and individual memory. In tandem with (and to some degree inspiring) such critiques, a new breed of countermonument has tried both to underline and to circumvent such abstraction" (1096).

8. Casey emphasizes the otherness of physical memorials when he writes that commemoration is a "highly mediated affair" that "involves a quite significant component of otherness at every turn" (218). He also suggests that it is "the very hardness and hardiness of granite or marble" that "concretize[s] the wish to continue honoring into the quite indefinite future—and thus, by warding off the ravages of time, to make commemoration possible at any (at least foreseeable) time" (226).

9. http://www.virtualmemorials.com/servlet.GuestBook (Dec. 2001). As of July 2003, the page has been removed.

10. http://daze.com/condolences/ (Dec. 2001). As of July 2003, the page has changed.

11. Sharon Ankerson. http://www.virtual-memorials.com/servlet/GuestBook (July 2003).

12. http://www.virtual-memorials.com/servlet.ViewMemorials?memid = 24822&pageno = 1 (July 2003).

16. http://virtual-memorials.com/articles/abcnews.html (July 2003).

14. http://virtual-memorials.com/articles/abcnews.html (July 2003).

15. http://virtual-memorials.com/servlet.ViewMemorials?memid = 5241 & pageno = 1 (July 2003).

16. Gary Delaplane. http://virtual-memorials.com/servlet.ViewMemorials? memid = 24822&pageno = 3 (July 2003).

17. The Internet plays a loaded role in the Columbine debate other than the memorials that sprang up out of it. It is well known that the killers, Dylan Klebold and Eric Harris, had their own websites, which were later examined by investigators. Harris and Klebold themselves seemed to have relished the idea of their own posthumous celebrity status: in the tapes they made, they talk of the movie that will be made out of the killings, imagine which director might direct the movie, and talk of wanting to live forever, as ghosts that haunt the survivors. See "The Columbine Tapes."

18. Money made from the book goes to the Cassie Bernal Foundation, which has funded an orphanage set up in Honduras, run by Christian missionaries from North Carolina.

19. On September 23, 1999, Dave Cullen's "Inside the Columbine High Investigation" raised the possibility, due to new eyewitness reports and testimonies, that it was not Cassie that was asked the question, but someone else. As of July, 2003, Zoba's article is no longer linked to Cassie Bernall's website.

20. http://www.cassiebernall/com/cassie_bernall_Parents.htm (Dec. 2001). As of July 2003, the quote is no longer available on this site.

21. This photograph is the same one that appears on the book *Rachel's Tears*, written by Rachel's father Darrell Scott.

22. http://www.racheljoyscott.com/ (Dec. 2001). As of July 2003, the web memorial has been redesigned.

23. Harris wrote, "We will hijack a hell of a lot of bombs and crash a plane into NYC." See "Columbine Killer"15.

24. The webpage's new design and layout still emphasizes Christianity and Rachel's faith as did the previous one.

25. www.racheljoyscott.com/rachelslegacy.htm (July 2003).

26. Although all three memorials discussed here have links to each other, Daniel Mauser's is the only one to show photographs of all the other victims, most of whom don't have their own web memorials.

27. The slurs that appear on Mauser's memorial are reminiscent of the way in which the Harburg memorial was vandalized. Illegible scrawls, hearts, stars of David, funny faces, and Swastikas all appeared on it. It thus became a social mirror, reminding the community not only of what had happened, but of how they now responded to the memory of that past.

WORKS CITED

Casey, Edward. *Remembering: A Phenomenological Study*. Bloomington: Indiana UP, 1987.

Cassiebernall.com. Jan. 2002 and July 2003 http://cassiebernall.com/.

"The Columbine Tapes." *Time* 20 Dec. 1999: 40-51.

"Columbine Killer Mapped Out His Tactics in a Diary." *The Independent* 8 Dec. 2001: 15.

Crane, Susan. Introduction. *Museums and Memory*. Ed. Susan Crane. Stanford: Stanford UP, 2000. 1-13.

Cullen, Dave. "Inside the Columbine High Investigation: Everything You Know about the Littleton Killings is Wrong." *Salon.com* 23 Sept. 1999.

Danielmauser.com. Jan. 2002 and July 2003 http://www.danielmauser.com.

Daze.com. Jan. 2002 and July 2003 http://daze.com.

Handelman, Don, and Lea Shamgar-Handelman. "The Presence of the Dead: Memorials of National Death in Israel." *Suomen Antropologi* 16.4 (1991): 3-17.

Huyssen, Andreas. "Monument and Memory in a Postmodern Age." *Yale Journal of Criticism* 6 (1993): 249-61.

Linenthal, Edward T. "Locating Holocaust Memory: The United States Holocaust Memorial Museum." *American Sacred Space*. Ed. David Chidester and Edward T. Linenthal. 220-61.

———. "Memory, Memorial, and the Oklahoma City Bombing." *Chronicle of Higher Education* 6 Nov. 1998: B4+.

Nora, Pierre. "Between Memory and History: Les Lieux de Mémoire." *History and Memory in African-American Culture*. Ed. Geneviève Fabre and Robert O'Meally. New York: Oxford UP, 1994. 284-300.

Porter, David. Introduction. *Internet Culture*. New York: Routledge, 1997. xi-xviii.

Poster, Mark. *The Mode of Information: Poststructuralism and Social Context*. Chicago: U of Chicago P, 1990.

Pray for President Bush. Dec. 2001 and July 2003. http://www.prayforgeorgewbush.com/pages/236635/index.htm.

Racheljoyscott.com. Jan. 2002 and July 2003 http://www.racheljoyscott.com.

Stream, Carol. "Church, State, and Columbine." *Christianity Today*. 2 April 2001: 54-59.

Trumpener, Katie. "Memories Carved in Granite: Great War Memorials and Everyday Life." *PMLA* 115 (2000): 1096-103.

Virtualmemorials.com. Jan. 2002 and July 2003 http://www.virtualmemorials.com.

Wagner-Pacifici, Robin, and Barry Schwartz. "The Vietnam Veterans Memorial: Commemorating a Difficult Past." *AJS* 97 (1991): 376-420.

Wallace, Susan. "Voices of Columbine." April 16, 2000. Dec. 2001 and July 2003 http://www.cassiebernall.org/legacy.htm.

Young, James. *The Texture of Memory: Holocaust Memorials and Meaning*. New Haven: Yale UP, 1993.

Žižek, Slavoj. *The Plague of Fantasies*. London: Verso, 1997.

Zoba, Wendy Murray. "Cassie Said Yes, They Say No." *Christianity Today* 6 December 1999: 77-78.

10
NETWORK THEORY AND LIFE ON THE INTERNET

John Johnston

In the early 1990s, a new sense of network connectedness found popular expression in John Guare's play, *Six Degrees of Separation*, which later became a successful movie. The character Ouisa explains the general idea: "Everybody on this planet is separated by only six other people. . . . The president of the United States. A gondolier in Venice. . . . It's not just the big names. It's anyone. A native in the rain forest. A Tierra del Fuegan. An Eskimo. I am bound to everybody on this planet by a trail of six people" (14). In the mid-1990s, a mini-phenomenon popular with college students known as the "Kevin Bacon Game" conveyed a similar sense. It started when three students appeared on a television talk show and demonstrated that Kevin Bacon could be linked through appearances in movies with every known Hollywood actor or actress, either living or dead.[1] For example, Bacon and Mike Myers have never appeared together in the same movie, but Myers can be linked to Robert Wagner through *The Spy Who Shagged Me* and Wagner appeared with Bacon in *Wild Things*. Myers thus has a "Bacon Number Two," indicating two degrees of separation from Bacon himself. After two computer scientists, using the vast Internet Movie Database, set up a Website where anyone could play the game, mathematicians could easily prove that no Hollywood actor or actress was ever more than four degrees removed from Bacon.[2] In the late 1990s, as network theory began to assume a new validity and importance in contemporary science, these two popular examples were often cited as instances of "small-world" networks. In fact, a number of scientists had begun to wonder if the Internet, despite its mushrooming, anarchic growth, also functioned as a small-world. But it turns out that the Internet possesses a stranger and more unexpected uniqueness.

Historically, the first recognition of the existence of "small worlds" followed from an experiment performed in the 1960s by the social psychologist Stanley Milgram. In order to determine how many social links separated mid-westerners living in Kansas and Nebraska from a stockbroker friend in Boston, Milgram asked a random selection from among the former to send a letter to the stockbroker, but without giving his Boston address. Those selected were not to try to mail the letter directly, but to send it to someone they knew personally who would be "closer" to or more likely to know the stockbroker. The recipients of these letters were then to do the same. Surprisingly, not only did most of the letters eventually reach the stockbroker, but the average number of mailings or intermediary links was six—hence the phrase, "six degrees of separation." In an article summarizing his experiment, Milgram referred to this unexpected shortness of distance between people as "the small world problem" (60–67).

Before Milgram's research, only mathematicians had exhibited a serious interest in networks. The great Swiss-born mathematician Leonhard Euler initiated this interest in 1736 when he solved a conundrum popular among the citizens of Konigsberg. The city was renowned for its seven bridges that connected a small island with both sides of the Pregel River, and for centuries the citizens had argued over whether one could traverse all seven without passing over any one bridge twice. Treating the bridges as a "graph" with hidden properties, Euler proved that it could not be done, thereby establishing the first theorem in "graph theory."[3] (In mathematics, a "graph" is defined as a collection of points called "nodes" or "vertices" connected by "links" or "edges.") Subsequently, and for the next two hundred years or so, mathematical interest focused primarily on regular graphs, in which each node has exactly the same number of links.

In the twentieth century, however, two Hungarian mathematicians, Paul Erdos and Alfred Renyi, invented the formal theory of random graphs, where the network of nodes and links is connected in a completely random fashion.[4] In contrast to classical or regular graphs, where links are always connected to neighboring nodes and the network therefore exhibits a high degree of local "clustering," random graphs often exhibit a high degree of connectivity across the network as a whole. In fact, as Erdos and Renyi discovered, only a very small number of randomly placed links is always sufficient to tie together a fragmented random network into a completely connected whole. To illustrate, let us suppose that in a remote area there are fifty small towns without any connecting roads. The government decides to build a network of roads, but only has a limited budget. How can the towns be connected with the least number of roads? Erdos and Renyi showed that about ninety-eight roads randomly placed would ensure that a great majority of the towns would be linked together.[5] While at first ninety-eight may seem like a large number, it is a little less than two roads per town. (Contrarily, if each town were connected to each of the other forty-nine towns, the total number of roads would be 1,225.) Although certainly counterintuitive, linking the towns randomly turns out to be a very efficient method.

Initially, random network theory seemed to explain Milgram's "small world problem"; that is, it seemed to account for why large numbers of randomly selected people could be connected by so few links. This was because random networks exhibit the high connectivity of "small worlds": since there are always a number of "shortcut" links connecting distant parts of the network, the average number of links or degrees of separation between any two nodes is never very large. The problem with this explanation, however, is that while our social networks may contain random factors, they are certainly not random. In 1973, taking up Milgram's idea of "the small world problem," the sociologist Mark Granovetter was able to shed considerable light on the nature of social networks by distinguishing between strong ties and weak ties.[6] Designating the individual subject as Ego, Granovetter reasoned that Ego's social world would be constituted by two kinds of connections: strong ties among family relations, close friends and co-workers, and weak ties among acquaintances and others whom Ego would recognize but not be closely attached to. These weak ties, moreover, are extremely important for society as a whole. As Granovetter puts it,

> Each of [Ego's] acquaintances [. . .] is likely to have close friends in his or her own right and therefore to be enmeshed in a closely knit clump of social structure, but one different from Ego's. The weak tie between Ego and his or her acquaintance, therefore, becomes not merely a trivial acquaintance tie, but rather a crucial bridge between the two densely knit clumps of close friends. [. . .] These clumps would not, in fact, be connected to one another at all were it not for the existence of weak ties. (qtd. in Buchanan 46)

Weak ties, therefore, form essential bridges among "closely knit clump[s] of social structure." Indeed, without these bridges there would be no social network, only isolated islands of small, densely connected groups. Not surprisingly, the "strength of weak ties," as Granovetter entitled his ground-breaking paper, turns out to be essential for most forms of social communication. Fads, rumors, and most sexually transmitted diseases spread rapidly because of weak links. In fact, Granovetter's initial research revealed a simple but compelling instance of the efficacy of weak ties. While still a graduate student in the late 1960s he asked dozens of people in a working class Boston suburb how they had obtained their current jobs. Overwhelmingly, he got the same reply: through an acquaintance (that is, a weak tie). In the 1980s, a dim but widespread recognition of the importance of weak ties surfaced in the popular buzz-term "networking."

Until recently, contributions to the nascent science of network theory seemed to accumulate slowly, almost haphazardly, and from a wide diversity of researchers. Not unexpectedly, some of the early ideas were ignored and only rediscovered later. Perhaps the most striking example is Paul Baran's proposal in the early 1960s of a new kind of communication network.[7] Immediately upon his arrived at the Rand Corporation, he was handed the task of developing a communications network that could survive a nuclear attack from the Soviet Union. Against all odds, he actually came up with a viable plan. Baran realized that of the three possible architectures—centralized, decentralized and distributed structures—only the mesh-like architecture of a distributed network would not be easily vulnerable. Whereas destruction of the central hub or hubs in the centralized and decentralized structures would cripple the network, in the distributed structure, where messages can travel along any number of routes, the network would continue to function even if large numbers of nodes were destroyed. However, in addition to the complete novelty of the proposal there was a technical obstacle. At the time, the existing communication network was an analog system controlled by the AT&T monopoly. Since Baran's proposal called for messages to be sent in small packets of uniform size that could travel independently along any route in the network, a changeover to a digital system would be necessary. But this was not something that "Ma Bell" was prepared to do, so Baran's proposal was buried. Many years later, when ARPA (the federal government's Advanced Research Projects Agency) began to design ARPANET, which would become a prototype of the Internet, it unknowingly adopted a version of Baran's original idea.

Today, the Internet stands as one of the most striking instances of a human-made network. Its singularity resides not simply in its spectacular growth—it has doubled in size every year for the past ten years—but also, and less familiarly, in the mystery of its underlying dynamics and peculiar topography. These properties only began to reveal themselves in the late 1990s, when initial efforts were made to measure its size and map its expanse. In several crucial instances, these efforts converged with further developments of network theory. Indeed, that scientists soon found the Internet to be a privileged object of study—interesting in and for itself, as if it were some new kind of ecology—is a no less singular fact to which we will later return.

Given the almost anarchic freedom and uncontrolled growth of the Internet, researchers first assumed that it constituted a random network. From Erdos and Renyi's work it was known that in random networks most nodes possess about the same average number of links; that is, only a small number of nodes will have either very few or a high number of links. In mathematical terms, this means that the probability calculations for the number of links per node in a random network is described by a Poisson distribution or Bell curve. (In contrast to classical networks, where the properties can be determined directly, in random networks they can only be specified statistically.) The "peak" of the curve and the distinctive cutoff points on either side will indicate its characteristic scale. However, it was soon discovered that the average number of links per node for both Internet routers and website pages do not follow a Bell curve but exhibit a power law distribution.[8] This was a startling discovery, since it suggested that some kind of heretofore unsuspected law was governing the shape or topography of the Internet as a whole. Indeed, that both the virtual network formed by website pages and their connecting links and the physical network (that is, the material infrastructure) formed by the routers (the "nodes") connected by the telephone lines and fiber optic cables (the "links") exhibited a power law distribution in their link-node structure made this conclusion unavoidable.

What makes the power law distribution so different? Albert-Laszlo Barabási, one of the researchers who discovered this feature of the Internet, illustrates the difference between a Bell curve and power law distribution by comparing the US highway system with the US air traffic system. As a network, the highway system is basically random, with most cities being served by roughly the same number of highways.[9] This means that there are almost no cities that have either a large number or only one or two highways, and consequently that the highway system has a characteristic scale, indicated by the peak of a Bell curve. The air traffic system, in contrast, is not random but exhibits a certain kind of order: a large number of small airports are connected to each other via a few major hubs, which in turn are connected to many airports both large and small. As a consequence, there is a much wider range in the distribution of links and nodes.

In addition to a wide range of differences in links per node, the power law distribution exhibits no cutoff points at the extremes. For this reason it is said to be *scale-free*. Again, a simple example may prove helpful. If we measured the height of a thousand American males of age forty we would find that the measurements fall under a Bell curve, with the average height around 5'9." Conversely, we would not find that their heights ranged between wide extremes, say between one or two and 30 or 40 feet, with the average falling just short of 6 feet. In other words, we would find that there is a characteristic scale in height for men of age forty, and that typical examples would not vary greatly from the average. Contrarily, for a scale-free network like the air traffic system the "average" number of links per node is not a useful notion. Since most nodes are very poorly connected while a few others are highly connected, the average number gives a very misleading picture of the distribution.

Barabási and his research group discovered that the Internet is just such a scale-free network, with the number of links to nodes ranging widely from hundreds of thousands to one or two. Thus, the average number of links to nodes is not a useful piece of information, and typical examples vary enormously from the average. While to the layperson this network may appear to be more random or chaotic (that is, less predicable) than the so-called random network, to the mathematically trained eye a scale-free power law distribution indicates the presence of a hidden order. Given that the World Wide Web is not a random but a scale-free network, the question naturally posed itself: does it also exhibit "small world properties"? Is the Web, in short, a small world? To determine the size and linking structure of the Web, Barabási and his research group deployed a software "robot" that wandered from site to site, systematically following and mapping all of the links emanating from each site. They also drew on the results of other research groups. What they discovered was that the Web (in the late 1990s) was comprised of roughly a billion nodes (that is, Web pages), and that it exhibited about nineteen degrees of freedom. On average, then, any Website was about nineteen clicks away from any other. Although nineteen is much larger than the now proverbial six degrees of freedom, for Barabási and his group it qualified the Web for small-world status. These calculations implied, furthermore, that even if the Web should continue to grow at its present dizzying rate, this number would only increase incrementally.

On the other hand, the group also discovered that this number—and the concomitant assumption that the Web is a small world—is highly misleading, since, given the Web's tremendous size, the number nineteen (though large) suggests that the Web is easily navigable. They found, contrarily, that starting from any given Web page we can only reach about twenty-four percent of all documents. And this is not simply because search engines are not yet very efficient. Several factors explain why much of the Web remains invisible. For one thing, it is much larger and is growing much faster than anyone had previously thought. Furthermore, since the number of links to any particular site is what determines its "visibility," sites with only a few links are obscure and hard to find, and vastly overshadowed by large "hubs" like amazon.com which possess hundreds of thousands of links. But size and link structure are not the only factors inhibiting visibility. There are also topographical obstacles that result from the technical fact that most Web links are directed, meaning that these links only work in one direction and that to return to an originating or initial site requires that we follow a different route. Indeed, the presence of these directed links has a large and fateful consequence, fragmenting the Web as a whole into what Barabási calls three major continents and a fourth made up of islands and tendrils, all of which severely restricts movement within and across the Web. As Barabási indicates, once you're in the "central core" you can't get back to the "in continent," and once you're in the "out continent," you can't get back to the central core.

Other peculiarities of the Web have also come to light as a result of the growing scientific interest in network theory. Perhaps, the most interesting is the sense in which it constitutes a complex evolving system, making it more like a biological cell than a computer chip, as Barabási puts it (*Linked* 149). Before we can explore this aspect of the Web's vitality, however, we must return to the larger picture that was beginning to form in the late 1990s, as scattered but growing discoveries began to "jell" or come together and network theory emerged as an officially recognized scientific pursuit. There is general agreement that if a single event could be said to have both catalyzed and intensified widespread scientific interest in the dynamics of networks, it was the discovery of the underlying mathematical structure of the "small world." Recall from the discussion above that with Milgram's and Granovetter's research a distinctive third type of network had clearly emerged. That is, the "small world" network was neither the regular network of traditional graph theory nor one of Erdos and Renyi's random networks. But what exactly were its distinctive mathematical properties? In 1998, the applied mathematicians Duncan Watts and Steven Strogatz published a three-page article in the prestigious science journal *Nature* that supplied an answer.[10]

Basically, Watts and Strogatz started with a regular lattice in which each node is connected to four of its adjacent neighbors—two on each side. Since each node is directly connected to its close neighbors, this kind of network exhibits a high degree of local "clustering." Then, by simply replacing a few of these near-neighbor links with links to randomly selected nodes throughout the network, they produced an entirely different kind of network, one that exhibited small world properties. Continuing this process, which they called "random rewiring," they eventually produced a completely random network. To compare the behavior of these three types of network, Watts and Strogatz made two kinds of measurements: the characteristic path length and the clustering coefficient. The characteristic path length is simply the number of edges (links) in the shortest path between two vertices (nodes) averaged over all pairs of vertices. The clustering coefficient was found by taking each vertice, counting the largest number of possible edges, then dividing by the actual number of edges, then calculating the average value of this ratio over the entire network. These two measurements yielded the sought after mathematical properties of the small-world architecture, specifically, a distinctive combination of high clustering with short characteristic path length. Taken together, these two indicators confirmed in quantitative terms what might be called the visual "signature" of a small-world network: the combination of local clustering (evident in the regular lattice) with shortcuts across the network (evident in random networks). Compared to regular lattices, then, small-world networks have shortcuts; compared to random networks, they have local clustering and are "sparse."

To test their mathematical model, Watts and Strogatz measured the characteristic path length and clustering coefficient for three real-world networks: the collaboration graph of Hollywood film actors (the Kevin Bacon game), the electrical power grid of the western United States, and the neural network of the nematode worm *C. elegans*. (Detailed databases and mappings for all three were already available.) In each instance, the measurements were also compared with those from random networks with the same number of vertices and average number of edges per vertice. The results showed that all three real-world networks clearly exhibit the mathematical properties of small-worlds, thus suggesting to Watts and Strogatz that the small-worlds phenomenon is "not merely a curiosity of social networks, nor an artifact of an idealized model—it is probably generic for many large, spare networks found in nature" (441).

Finally, Watts and Strogatz considered the functional significance of small-world connectivity for dynamic systems, using a deliberately simplified model of the spread of infectious diseases as their test case. Given the two features that distinguish small worlds (clustering and shortcuts), it is hardly surprising to discover that infectious diseases spread more easily and more quickly in this kind of network. However, and in contrast to other network models of disease spreading, their work shows how a significant increase of spreading is a structural feature, rather than a simple matter of topography (number of links). In the spread of sexually transmitted diseases, for example, it is not simply the number of concurrent partners that individuals have that determines the increase (a fact obvious to the individual), but the structural transitions that lead to small worlds, which is too subtle for the individual to observe. In fact, what is most alarming in this context is how few shortcuts are needed to make the world "small."

In their conclusion, Watts and Strogatz point out that small-world connectivity has implications for other dynamic systems as well: for enhanced signal-propagation in cellular automata, computational power in iterations of "Prisoner's Dilemma," and the synchronization of small-world networks of coupled phase oscillators. As they also add, the observed synchronization of widely separated neurons in the visual cortex may even suggest that the brain itself has a small-world architecture. Indeed, it is not at all difficult to see why this architecture is to be found in many natural and artificial systems: by combining a relatively high degree of local clustering with the ability to communicate across the network, it offers an extremely economic solution to contradictory demands.

That small-world networks might have significant implications for dynamic systems was certain to catch the attention of scientists at the Santa Fe Institute, a renowned center devoted to the study of complex dynamic systems.[11] Little surprise, then, that in August 2000 the Institute hosted a conference on "Complex Interactive Networks" in order to define the central issues. As expected, the discussion focused on the peculiar and often non-intuitive properties of networks, both natural and artificial, random and constructed: how they grow and how that growth influences their patterns of connection, their cohesion, functionality, and likelihood of collapse.[12] The actual research considered spanned a wide range of scientific disciplines, from mathematics and computer science to molecular biology, neuroscience, entomology, and social anthropology. This too was hardly surprising, given that research on complex dynamic systems is interdisciplinary by nature and that a good deal of research on networks had already been done at the Institute.

Apart from its official acknowledgment of the scientific importance of network theory, the most notable aspect of the conference for our purposes was the discussion of the Erdos-Renyi and Watts-Strogatz network models, and more specifically the realization that neither model successfully reproduces the patterns of connectivity that characterize the Internet. Both of these models are static, since the number of nodes and the structure of their links remain fixed. But the World Wide Web was (and still is) growing, and thus any account of its structure will have to assume a dynamic model. Once again, it was Albert-Lazslo Barabási and his research group who first proposed a dynamical model of the Web.[13] Like other research groups, they had observed that the Web's link-node structure exhibits a power law, and they were the first to emphasize its scale-free character. They had also noticed that the Web is characterized—despite its "democratic" appearances—by a number of "hubs" with a large number of links. Sites like Amazon.com is an obvious example, but there are many others. For Barabási and his group, these "hubs" or "connector" links turn out to be fundamental for understanding the dynamics of the Web's growth. They realized that as new sites are added, their new links are not picked randomly. Rather, since links are chosen to enhance connectivity and visibility, most likely they will be added to large and well known sites. This tendency, which they call "preferential attachment," leads inevitably to an effect very familiar in dynamical systems theory and encapsulated in various phrases like "them that has, get more," "the rich get richer" and "winner take all." In academia, it explains the frequently repeated remark that most of the grants seem to go to the same faculty, while their often more talented and original colleagues get very little support at all. As concerns the World Wide Web's continual growth, this effect leads to greater "hub" formation and an overall increase of links to these preferred "connector" hubs. Them that already has, get more.

In essence, Barabási and his group established that the Web is governed by two laws: growth and preferential attachment. In their 1999 *Science* article, they proposed a network model that incorporated both. Growth, and the temporal structure it assumes, gives a distinct advantage to older sites, since over time they are much more likely to acquire new links. Growth alone, however, does not generate the power laws, which seem to require the presence of hubs. But when preferential attachment is added to growth a multiplication effect occurs, since now the older, more numerously linked sites are more likely to acquire the lion's share of the new links as well. The theory was confirmed when Barabási's former student Reka Albert translated growth and preferential attachment into simple algorithms and then, using computer simulations, was able to demonstrate that together they generated the mysterious power laws. As a consequence, and for the first time, a new model of a scale-free network could account for both the hubs and the power laws evident in a real network.

This was a particularly exciting discovery for physicists, for it established clear connections between network theory and current research in fields like phase transition in molecular states, percolation theory, and research in metabolic pathways, to name just a few. In a recent summary article,[14] Albert and Barabási trace some of these parallels, but of more direct relevance here is Barabási's suggestion in *Linked* that the scale-free model embodies "a new modeling philosophy," since it both creates "a set of opposites: *static* versus *growing*, *random* versus *scale-free*, *structure* versus *evolution*," and necessitates a shift "from describing the topology to understanding the mechanisms that shape network evolution" (90, 91). Needless to add, the shift toward understanding the self-organizing principles of networks as a function of their growth and evolution promises to lead to a deeper understanding of many complex dynamic systems. Two examples of this shift, both developed in Internet research, are worth mentioning briefly.[15]

By applying the kind of probability equations that physicists use in statistical mechanics for calculating molecular states, Barabási has attempted to define a "fitness function" for a website page. (He embarked on this research problem after wondering why the Google search engine had so easily beat out its competitors.) Essentially, this fitness function would measure a node's ability to compete for links at the expense of other nodes. While a node's "fitness" is apparently determined by factors like good or useful content and an attractive appearance, Barabási sought a measure that would reflect how each node's fitness evolves over time; more specifically, he was interested in how fit nodes could join the network at some later time and connect to many more links than older but less fit nodes. Thus, he sought a measure that would refine and complicate the growth and preferential attachment model proposed earlier, by providing a further quantitative measure of the Web's dynamic, evolutionary growth.[16]

The second example concerns the Web's vulnerability, particularly to hacker attack with computer viruses and worms. Given the Web's greatly increased importance to the economy, this is no small concern. So far, of course, the Internet has proven to be extremely resilient to both computer viruses and hardware failures. This robust quality is largely due to the nature of a scale-free network. Indeed, using computer simulations Barabási discovered that removing as many as 80% of the nodes—if selected randomly—would not break down a scale-free network. However, he also discovered that if hubs or connector nodes are targeted systematically, the network becomes highly vulnerable. This Achilles heel, as Barabási calls it, is another consequence of a scale-free architecture. The very same feature that makes the Web relatively impermeable to random failures or attacks also accounts for its vulnerability to selective attack.

While this scenario now fits with our expectations, for computer viruses the situation is surprisingly unexpected. Epidemiologists have long studied the spread of disease using diffusion models with a specifically defined critical threshold. Simply put, the threshold indicates the disease's contagion potential: diseases that are contagious below a certain critical threshold soon die out, while those above it spread exponentially until eventually reaching most of the population. A research group working in Barcelona and Trieste has discovered that in scale-free networks the critical threshold is zero.[17] This means that in a scale-free network like the Web even weak computer viruses (those with a low critical threshold) will spread and persist indefinitely. As Barabási points how, this discovery explains why the most destructive computer virus seen thus far, the so-called "Love Bug," remains the seventh most frequently encountered virus well over a year after its supposed eradication.

From their first appearance, computer viruses were deemed to be a new and undesirable form of "artificial life." This was especially true among early researchers in the science of Artificial Life, who were anxious that their new discipline not be associated with the dangerous and destructive practice of anarchic hackers. Yet, the proliferation of new digital life forms—both the *bonafide* productions/experiments of A-Life scientists and the uncontained forms released by hackers—could not help but produce the impression that a whole new realm of artificial life is burgeoning around us.

Not surprisingly, then, in her research on a viable computer immune system the computer scientist Stephanie Forrest has come to conceive of the world of computers as having many of the properties of a living ecosystem, populated with "computers, with people, software, data, and programs."[18] In "Principles of a Computer Immune System,"[19] Forrest lists some of the organizing principles of a biological immune system—autonomy, adaptability, and a dynamically changing coverage—that must be incorporated as design principles if a computer immune system is to function. However, if the objective is "to design systems based on direct mappings between system components and current computer system architectures," then the latter will have to be radically modified (79). One possible architecture, she suggests, would be something like an equivalent "lymphocyte process" comprised of lots of little programs that would query other programs and system functions to determine whether they were behaving normally or not. But they would also have to monitor each other, "ameliorating the dangers of rogue self-replicating mobile lymphocytes" (80). Just how feasible this approach will turn out to be remains uncertain, and Forrest is rightly cautious. In fact, she is acutely aware of the limitations of "imitating biology," since biological organism and human-made computers have very different methods and objectives.

Even with this caution in mind, however, it is not difficult to imagine applying the idea of an immune system to the Internet itself, particularly since it so clearly exhibits many of the properties of a living ecosystem. Although a humanly constructed artifact of contemporary technology, its laws of internal structure, growth and evolution seem more like those of a natural living form. To a large part, moreover, its self-organizing properties continue to function unperturbed by and completely independent of human intention. In one sense, the Internet is like a vast beehive or ant colony: by tirelessly adding new websites, establishing new links and repairing mechanical breakdowns, human users are building and maintaining a structure whose global dynamic behavior everywhere exceeds their individual desires and awareness. In itself, of course, this is not so unusual—human beings have always been tangled in many similar systems. But the Internet is unique in at least two fundamental respects. At the level of the individual user, it answers to a number of uses or functions, as reflected in the heterogeneous metaphors we use to describe it. The poet Stephanie Strictland succinctly highlights this aspect when she describes the Web as

> an enormous structure, almost biological in the way it communicates and propagates by proliferating links. The electronic space, often called cyberspace, has some very unusual qualities, to judge by pre-electronic categories. It is characterized as tidal sea, web, sky, and solid. Thus, people surf it, send out web-crawlers to explore it, gophers to tunnel through it, engines to mine data from it, and they fly through and above it in game simulations. They establish "home" pages in it, as though it were rooted, although at their own location distance has disappeared—New Zealand, New York, St. Paul, equally present, and equally speedily present.[20]

When we move beyond the level of the individual user and consider the use of the Internet by corporations and businesses, institutions and governments, we are struck more forcefully by the vast databases and sophisticated communications software that enable workers and professionals of almost every stamp to access, exchange, and process information on a global scale. This consideration leads directly to the second aspect of the Internet's uniqueness: as a network it contains not only more information but more "intelligence" than anything ever constructed in human history.

Not surprisingly, it is this second aspect that has generated speculation about the Internet as an inchoate "global brain." While much—perhaps most—of this speculation is wild-eyed and mystical, there are good reasons why it cannot be simply dismissed out of hand. Many highly reputable scientists like Barabási have noted that the earth is literally enfolded in "millions of measuring devices, including cameras, microphones, thermostats and temperature gauges, light and traffic sensors, and pollution detectors [that are] feeding information into increasingly fast and sophisticated computers" (*Linked* 158). The number of these computers as well as Internet-connected cell phones is increasing almost exponentially. Now, not only are these devices rapidly increasing in number but for the first time they are feeding information into a single integrated system. As a result, on a material level the Internet is already the site of a vast network of connected sensors and processors. The possibility that it might self-organize into a computer functioning unpredictably and independently of human supervision is therefore not altogether a science fiction fantasy.[21] The further possibility that such a networked computer might become self-aware is far less likely, though not merely a baseless projection. What this scenario reflects, however naively, is an inchoate sense that if the Internet should become a computer it will not be like the one on our desks, which instantiates and functions according to strict mechanical laws, but rather like the human brain, which, while never violating those same laws, has grown and re-wired itself in response to both internal principles of organization and function and adaptive necessities arising from its interface with the environment.

All speculation aside, there can be no doubt that the Internet possesses a life of its own, and may be evolving into a complex adaptive system. This "life," as we have seen, has come about as a direct result of its structure, topology and dynamical growth, in turn the product of the ongoing activities of its human users. But the Internet as a new type of information ecology and environment also supports a wide range of digital life forms. In addition to the computer viruses and data-mining software already mentioned, all kinds of knobots and "smart" autonomous agents increasingly roam the Net, while many AI-inspired programs like "Eliza," "Ramona" and various chatterbots reside at numerous sites. Perhaps the most intellectually interesting of digital life forms on the Net, however, are those produced by Artificial Life scientists. Tom Ray's Internet version of Tierra provides a striking case in point.

In Ray's original version, Tierra was a virtual world in which digital organisms (blocks of code) could replicate, mutate and compete for processor time and memory in a laptop computer. Its resounding success—widely recognized by A-Life scientists and evolutionary biologists alike—eventually led Ray to the idea of relocating it on the Internet, where it could take advantage of the fact that at any given time there are thousands of "idling" machines on the network that could provide spare CPU cycles. The objective, Ray states in his initial proposal, would be "to set off a digital analog to the Cambrian explosion of diversity, in which multi-cellular digital organisms (parallel processes) will spontaneously increase in diversity and complexity." If successful, he continues, "this evolutionary process will allow us to find the natural form of parallel and distributed processes, and will generate complete digital information processes that fully utilize the capacities inherent in our parallel and networked hardware." The global computer network, because of its "size, topological complexity, and dynamically changing form and conditions" presents the ideal habitat for this kind of evolution.

In these propitious conditions, Ray hopes that individual digital organisms will evolve into multi-celled organisms, even if

> the cells that constitute an individual might be dispersed over the net. The remote cells might play a sensory function, relaying information about energy levels [i.e., availability of CPU time] around the net back to some "central nervous system" where the incoming sensory information can be processed and decisions made on appropriate actions. If there are some massively parallel machines participating in the virtual net, digital organisms may

choose to deploy their central nervous systems on these arrays of tightly coupled processors.

Furthermore, if anything like the Cambrian explosion occurs, then we should expect to see not only "better" forms of existing species of digital organisms but entirely new species or forms of "wild" software, "living free in the digital biodiversity reserve." Since the reserve will be in the public domain, anyone willing to make the effort will be able to observe and even "attempt to domesticate" these digital organisms. While domestication will present special problems, Ray foresees this as an area where private enterprise can get involved, especially since one obvious realm of application would be as "autonomous network agents."

After several years of operation, the Internet version of Tierra has not been as dramatically successful as the original version, mostly because of difficulties developing the parallel-processing software. Yet, there have been some rather remarkable results. At the Artificial Life VI conference in 1996, Ray and colleague Joseph Hart reported on the latest experiments, which they describe as follows: "Digital organisms essentially identical to those of the original Tierra experiment were provided with a sensory mechanism for obtaining data about conditions on other machines on the network; code for processing that data and making decisions based on the analysis, the digital equivalent of a nervous system; and effectors in the form of the ability to make directed movements between machines in the network" (303). Tests were then run to observe the migratory patterns of these new organisms. For the first few generations, these organisms would all "rush" to the "best-looking machines," as indeed their algorithms instructed them to do. The result was what Ray called "mob behavior." Over time, however, mutation and natural selection led to the evolution of a different algorithm, one that simply instructed the organism to avoid poor quality machines and consequently gave it an immense adaptive advantage over the others.

Ray's experiment is a clear instance of how the Internet itself has become a new kind of laboratory, one combining the features of both a natural environment and a new kind of technological space. In *Out of Control: The Rise of Neo-Biological Civilization*, Kevin Kelly argues that we have entered a new era in which human-made systems will increasingly take living organisms as their model; that is, in order to achieve a robust adaptability and autonomy our machines will have to be more biological in both their construction and modes of operation.[22] The Internet is a signal instance of a system that has assumed the shape and exhibited the dynamics of a quasi-biological organism, but quite independently of human intention and design. The Net's double status seems to have resulted from the fact that as a network it grows according to a pattern often found in nature, and that its electronic filaments support both human and digital forms of life. As a new kind of shared space or habitat, an ecology that is neither completely artificial nor natural, the Internet may be both the harbinger and testing ground of a new kind of environment in which technology continues evolution by other means.

NOTES

1. Barabási provides these details (*Linked* 58–59).
2. The two computer scientists, Glen Wasson and Brett Tjaden, called the website "The Oracle of Bacon": http://www.cs.virginia.edu/oracle/.
3. See West.
4. In 1959, 1960, and 1961 Erdos and Renyi published three now classic papers on random graph theory. In *Linked: The New Science of Networks*, Barabási provides a clear account of this work.
5. The illustration is taken from Buchanan 35–36.
6. Buchanan, on which I draw here, provides a useful summary of Granovetter's work.
7. See Baran for an indication.
8. A power law expresses some quantity a as a power of another quantity b, usually in inverse proportionality, as in $a = 1/b^n$. Power law distributions appear commonly in both nature and culture. When cities, earthquakes, and word distributions in natural languages are ranked according to size or frequency of appearance, they all exhibit power law distributions. For example, compared to the largest earthquake, the tenth largest earthquake appears one-hundred times more frequently.
9. The illustration is from *Linked* 71.
10. See Watts and Strogatz.
11. Walthrop provides a lively account of the formation of the Santa Fe Institute and its sponsorship of research on complex dynamical systems.
12. See Shalizi's summary account.
13. See Barabási and Albert.
14. See their "Statistical."
15. Both of these examples are drawn from Barabási "Physics."
16. Parallel efforts have been made to study the patterns of human users on the Web. Huberman sees in the Web "a veritable laboratory where one can study human behavior with a precision and on a scale never possible before" (16). What Huberman actually studies is the surfing patterns of people using the Web, which can easily be tracked using data already available. It turns out that there is "not only a law that describes the way we hop from link to link, but also an interesting insight into human behavior and the existence of a kind of economy of attention that guides our surfing" (42). Huberman arrives at these results by applying the methods of statistical physics to visitor data collected by large "portal sites" like Yahoo!, Excite, and MSN in order to measure the site's "stickiness"—that is, its capacity to keep the visitor at the site. Essentially, Huberman attempts to measure the "fluctuating value in the information" registered by visitors as they click from one Web page to another. Although each visitor surfs the Web according to a different mix of impulse, curiosity, and personal preference, statistical regularities emerge in the paths traced by thousands of visitors clicking through a specific site.

17. See Pastor-Satorras and Vespignani.
18. Quoted in King.
19. Presented at New Security Paradigms Workshop, Langdale, Cumbria UK, 1998.
20. See Strickland.
21. Barabási himself has explored how the Web might become a computer in experiments with "parasitic computing." See *Linked*, 156-99, for a brief discussion.

WORKS CITED

Barabási, Albert-Laszlo. *Linked: The New Science of Networks*. Cambridge: Perseus, 2002.

———. "The Physics of the Web." *physicsweb*. July, 2001. http://www.physicsweb.org/article/world/14/7/09 (28 July 2004)

———. "Statistical Mechanics of Complex Networks." *Review of Modern Physics* 74.1 (2002): 47-97.

Barabási, Albert-Laszlo, and R. Albert. "Emergence of Scaling in Random Networks." *Science* 286 (1999): 509-12.

Baran, Paul. "On Distributed Communications." 1964. http://www.rand.org/publications/RM/RM3420/index.html (28 July 2004).

Buchanan, Mark. *Nexus: Small Worlds and the Groundbreaking Science of Networks*. New York: Norton, 2002.

Guare, John. *Six Degrees of Separation*. New York: Vintage, 1990.

Huberman, Bernardo A. *The Laws of the Web: Patterns in the Ecology of Information*. Cambridge: MIT P, 2001

Kelly, Kelly. *Out of Control: The Rise of Neo-Biological Civilization*. Reading, MA: Addison-Wesley, 1994.

King, Lesley S. "Stephanie Forrest: Bushwacking Through the Computer Ecosystem." *SFI Bulletin* 15.1 (Spring 2000).

Milgram, Stanley. "The Small World Problem." *Psychology Today* 2 (1967): 60-67.

Pastor-Satorras, R., and Alessandro Vespignani. "Epidemic Spreading in Scale-Free Networks." *Physical Review Letters* 86 (2001): 3200–203.

Ray, Thomas S., and Joseph Hart." Evolution of Differentiated Multi-threaded Digital Organisms." *Artificial Life VI.* Ed. Christoph Adami, et al. Cambridge: MIT P, 1998.

Ray, Thomas S.. "A Proposal to Create a Network-Wide Biodiversity Reserve for Digital Organisms." 1995. http://www.hip.atr.co.jp/~ray/pubs/reserves/node1.html (28 July 2004; Internet Archive: http://web.archive.org/web/20001209041900/http://www.hip.atr.co.jp/~ray/pubs/reserves/node1.html

Shalizi, Cosma. "Growth, Form, Function, and Crashes." *SFI Bulletin* 15.2 (Fall 2000): 6–11.

Strickland, Stephanie. "Poetry in the Electronic Environment. 10 April 1997. http://www.altx.com/ebr/ebr5/strick.htm (28 July 2004).

Walthrop, M. Mitchell. *Complexity: The Emerging Science at the Edge of Order and Chaos.* New York: Simon, 1992.

Wasson, Glen, Patrick Reynolds, and Brett Tjaden. "The Oracle of Bacon." 1996. http://www.cs.virginia.edu/oracle/ (28 July 2004).

Watts, Duncan J., and Steven H. Strogatz. "Collective Dynamics of 'Small-World' Networks." *Nature* 393 (4 June 1998): 440–42.

West, D. B. *Introduction to Graph Theory.* Upper Saddle River, NJ.: Prentice, 1996.

CONTRIBUTORS

Jennifer L. Bay is an assistant professor of English at Purdue University. Her research focuses on gender in the workplace, new media and the body, and experiential learning. Her work has appeared in *College English* and *JAC*. She is at work on a manuscript about the rhetoric of maternity leave policies and the work/family conflict.

M.J. Braun is an assistant professor of rhetoric and composition at the University of West Florida. Her research focuses on the intersection of rhetoric, political-economy, and *logos*. Her most recent article, "The Prospects of Rhetoric in First Year Composition: Deliberative Discourse as a Vehicle of Change?" appears in the *Journal of Writing Program Administration* (2008).

Michelle Comstock is an assistant professor in the Department of English at the University of Colorado, Denver. She has published articles on gender, youth, and digital media literacy. Currently, she is collaborating on a book-length project regarding the privatization of literacy education and expertise.

Kristie S. Fleckenstein is an associate professor of English at Florida State University. Her research focuses on literacy issues related to imagery, embodiment, and technology. Her most recent book, which won the 2005 Conference on College Composition and Communication's Outstanding Book of the Year Award, is *Embodied Literacies: Imageword and a Poetics of Teaching* (Southern Illinois University Press, 2003).

Stanley D. Harrison is an assistant professor of English at the University of Massachusetts at Dartmouth. His Marxist research focuses on the subsumption of internetworked writing practices by capital. His most recent article is "Unconscious Writing in the Factory of the Social: A Class Theory of Negative, Allegorical Rhetoric," *JAC* (2007).

Byron Hawk is an associate professor of English at George Mason University. His primary research interests are histories and theories of composition and rhetoric and technology. His current book is *A Counter-History of Composition: Toward Methodologies of Complexity* (University of Pittsburgh Press, 2007).

John Johnston is professor of English and comparative literature at Emory University. His research focuses on contemporary literature, science and technology, and media theory. His most recent book, *The Allure of Machinic Life*, will be published by MIT Press in Spring 2008.

Ken S. McAllister is an associate professor of rhetoric at the University of Arizona and co-director of the Learning Games Initiative, an international and transdisciplinary research collective that studies, teaches with, and builds computer games in educational contexts.

Anthony J. Michel is an assistant professor of English at Avila University. His research focuses on the intersections of alternative rhetorics, social movements, and new media technologies. He has published articles in *JAC* and in various anthologies in rhetoric and composition.

Gary A. Olson is dean of the College of Arts and Sciences and professor of English at Illinois State University. He is editor (with Lynn Worsham) of *Postmodern Sophistry: Stanley Fish and the Critical Enterprise* (State U of New York Press, 2004).

Jim Ridolfo is a PhD candidate in the rhetoric and writing program at Michigan State University. His current dissertation research documents how media activists theorize the recompositon of their work. His recent work has appeared in *Kairos*, *Community Literacy Studies*, and *Pedagogy*.

Cynthia L. Selfe is Humanities Distinguished Professor in the Department of English at Ohio State University and co-editor, with Gail Hawisher, of *Computers and Composition: An International Journal*. Her research focuses on how literacy values and practices in digital environments shape—and have been shaped by—historic, economic, social, cultural, material, educational, and personal factors. Her most recent book is *Multimodal Composition: Resources for Teachers* (Hampton Press, 2007).

David Sheridan is an assistant professor in Michigan State University's Residential College in the Arts and Humanities. His research interests include digital and visual rhetoric (especially as these intersect with public rhetoric), alternative learning spaces for writing, and educational games. He is currently editing a collection focused on how writing centers can support students working on multimodal compositions.

Maya Socolovsky is an assistant professor of English at Iowa State University. Her research and publications focus on the narrativization of memory and history in U.S. Latino/a literature, and multi-ethnic literature. She is currently at work on a book project examining memory and place in U.S. Latino/a literature.

James J. Sosnoski is a professor of communication at the University of Illinois in Chicago. His research concerns learning through virtual reality experiences. His most recent book is *Configuring History: Teaching the Harlem Renaissance Through VR Cityscapes*.

Lynn Worsham is a professor of English at Illinois State University, where she serves as editor of *JAC*, the premier journal of theoretical scholarship on the intersections of rhetoric, writing, culture, and politics. Her most recent book is *The Politics of Possibility: Encountering the Radical Imagination*, edited with Gary A. Olson (Paradigm Press, 2007).

INDEX

abcnews.com, 193
Addison, Joanne, 51, 56
Advanced Research Projects Agency, 141, 210
Afghanistan, 139, 142
African Americans, 51
Agee, James, 128, 129, 137; *A Death in the Family*, 129
Air Force Office of Scientific Research, 141
Albert, Reka, 215, 220, 221; *Linked*, 215, 217, 220
Albrecht-Samarasinha, Leah Lilith,173
Allen, Nancy, 78, 88
Alliance for Community Media, 85, 88
Alternative Educational Environment's ASCEND Project, 141
Althusser, Louis, 98-99, 120, 132-33, 143; *Reading Capital*, 98
amazon.com, 197, 214
Amendment 22, 201
American Red Cross, 192
Andersen, Christopher, 192; *The Day Diana Died*, 192
Anderson, Sharon, 203
Anson, Chris, 96-97, 101, 105, 116-17, 120
Apple Computer, 87
Arendt, Hannah, 66
arête, 7, 11
Aristotle, 7, 10, 11, 17, 19-20, 71, 147, 153-54, 159; *DeAnima*, 20; *Nicomachean Ethics*, 17; *Poetics*, 20; *Rhetoric*, 5
Arlow, Jacob, 136, 143
Armstrong Laboratory, Crew Systems Directorate, 141; Human Resources Directorate, 141
Army Natick RD&E Center, 141
Army Research Laboratory, Human Research and Engineering Directorate, 141

Asen, Robert, 88
AT&T, 210
Athens, Greece, 8
Australian Film Commission, 179
avatar, 9, 32, 130-31

Balibar, Étienne, 98-99, 120; *Reading Capital*, 98
Ballif, Michelle, 19, 20
Bamboo Girl, 172
Barabási, Albert-Laszlo, 210-16, 220-21; *Linked*, 215, 217, 220
Baran, Paul, 209-10, 220-21
Barnes, Luann, 177, 185
Barnes and Noble, 197
Barson, James, 129
Barthes, Roland, 75, 76, 88, 153, 160; *The Semiotic Challenge*, 153
Bateson Gregory, 6, 7, 17, 19-20
Bateson, Mary Catherine, 10, 20
Batesonian cybernetics, 5
Baudrillard, Jean, 49, 50, 55, 148
Baumlin, James, 15, 21, 154, 160
Baumlin, Tita French, 15
Bay of Pigs, 150
Bay, Jennifer, *ix*
Bazerman, Charles, 84, 88
Benhabib, Seyla, 66
Belcher, Diane D., 51, 56
Bell curve, 210
Bellafante, Ginia, 180, 183
Bender, Walter, 128
Benthien, Claudia, 30, 38
Bergson, Henri-Louis, 158, 159
Berlin, James, 38-39, 96-97, 100, 105-08, 113, 120, 159-60; *Rhetorics, Poetics, Cultures*, 148
Bernall, Cassie, 197-99, 200, 203
Bernall Foundation, 203
Bernall, Misty, 198; *She Said Yes*, 197
Bernstein, Robin, 181, 183
Biesekcer, Barbara, 149-50, 159-60
Big Chill, The, 139
Bikini Kill, 170, 172-73, 177
Birmingham School, 80
Bitzer, Lloyd, 70, 88, 149-50, 160
Black Power Era, 139
Blair, Carole, 88
Blair, Jayson, 142

Blair, Kristine, 51, 56, 73
Blakesley, David, 38, 39; *Terministic Screen*, 38
Blockbuster Video, 131
Bloomfield, Harold H., 192; *How to Survive the Loss of a Love*, 192
bodies and writing, 25–40
Bolles, Edmund Blair, 21
Bolter, Jay David, 4, 178, 183
Booth, Wayne, 79, 88
Borders Books, 197
Boston, Massachusetts, 208, 209
Boundas, Constantin V., 35, 39
Bourdieu, Pierre, 175, 183
Brand, Stewart, 125, 143
Braun, M.J., *ix*
Brecht, Bertolt, 67
British Factory Act, 101
Brodkey, Linda, 169, 183
Brown, Janelle, 180, 183
Buchanan, Mark, 220–21
Buckley, Joanne, 51, 56
Buenos Aries, Argentina, 14
Bungle, Mr., 12
Burana, Lily, 172, 183
Burgin, Victor, 75, 89
Burke, Kenneth, 38–39, 129–30, 132,
    136, 143; *Rhetoric of Motives*, 136
Bush, George W., 199
Bust, 180
Bust Guide to the New Girl Order, 181

Calhoun, Craig, 66, 89
California State Senate, 110
Cambrian explosion, 218, 219
Campbell, Anne, 171, 183
Canada, 109
Carland, Tammy Rae, 171, 183
Carroll, Lewis, 3, 21
Casey, Edward, 188, 202, 204
Cato Institute, 112
Cayley, John, 4; *Indra's Net*, 4
Certeau, Michel de, 174, 183
Chan, Kristy, 172, 175, 177, 179, 183
chaos theory, 151
Chaput, Cathy, 120
Chelius, Terry, 201

Chickclick, 172, 183
Christ, Jesus, 196, 199
Christianity, 197, 199
Christianity Today, 197, 198
Churcher, Gavin, 87, 94
Civil War, 135
Clemson University, 88
Clinton-Gore administration, 113–15
Close, Chuck, 146
Clynes, Manfred E., 42, 56
Cold War, 113
Coleridge, Samuel Taylor, 157, 158
Colgrove, Melba, 192; *How to Survive the Loss of a Love*, 192
Colorado, 201
Columbine High School, 187–205
"Columbine Killer Mapped Out His Tactics in a Diary," 204
"Columbine Tapes, The," 204
Committee on Virtual Reality Research and Development, 126, 141
Computers and Composition, 182
Comstock, Michelle, *x*, 51, 56
Cooper, Marilyn, 74, 79, 92
Corbett, Edward P.J., 74, 75, 89
Cornyn, Alison, 77, 89
Cosby Show, The, 138
Crane, Susan, 191, 204
Cross, Rosie, 178, 179
Crowley, Sharon, 38, 39, 73, 93, 182–83
Cullen, David, 203, 204
CultPeeks, 126, 128, 133, 136–38, 140
Cumbria, United Kingdom, 221
Cupsize, 179
Cushman, Ellen, 89
CyberAngel, 142
Cyberbodies, 41–57
CyberEdge Information Services, 195

Danielmauser.com, 204
Daze.com, 204
Delaney, Ben, 195
Delaplane, Gary, 203
Deleuze, Gilles, 3, 21, 35, 39, 148, 155, 158–60, 189; *Spinoza*, 155, 159; *Thousand Plateaus*, 35
DeLuca, Kevin Michael, 64, 74, 89

democracy hope, 95-123
Department of Commerce, 101, 113, 119
Department of Defense, 142
Derrida, Jacques, 150; *Glas*, 150
Des Moines, Iowa, 201
Detroit, Michigan, 82, 131
DeVoss, Danielle, 62, 81, 89
Dewey, John, 107, 108
différance, 95
Dinwiddy, J.R., 95, 120
dissoi-logoi, 146-49, 153
dissoi-para-logoi, 153
Doane, Mary Ann, 172, 183
Downing, David, 96-97, 100, 105, 107-08, 113, 121
Dragga, Sam, 78, 89
Dreamweaver, 50
DSL, 50
Dublin, Ireland, 139
Duncombe, Stephen, 165, 167, 170, 174, 183
Durlach, Nathaniel, vii, 141, 143
Dyson, Anne Haas, 87, 89

Eastman, George, 65, 68
Eco, Umberto, 76, 79, 81, 89, 143
Educational Video Center, 82
Edwards, Paul N., 119, 121
Einstein, Albert, 98
Eliza, 218
Ellertson, Anthony, 62, 73, 89, 90
Emelye, 179, 183
Engberg-Pederson, Troels, 17, 21
Engels, Frederick, 48, 96, 100, 112, 121; *The Communist Manifesto*, 96, 100; *The German Ideology*, 48
England, 101
English studies, vii
enkrateia, 159
Enos, Richard Leo, 147, 160
Entertainment Weekly, 181
enthymeme, 152
Enzensberger, Hans, 67, 79, 90
Erdos, Paul, 208, 210, 212, 220
ethos, 3-20, 72, 154
Euler, Leonhard, 208
eunoia, 7, 11

Evan, Walker, 128
Excite, 220

Factory Act, 102
Faigley, Lester, 90
Faust, 126
Federal Reserve, 112
Feminist Press, 181
Feminist Studies, 181
feudalism, 98
Field of Empty Chairs, The, 190
Finnegan, Cara A., 62, 90
Firefox, 126
First World, 44, 46, 47
Fleckenstein, Kristie, ix
Florida, 14
fordism, 120
Forrest, Stephanie, 216
Fortune 500, 111
Foucault, Michel, 79, 98-99, 120-21, 154, 158-60; *The Use of Pleasure*, 154
Fraizer, Dan, 184
Fraser, Nancy, 66, 173, 182, 184
Frederick's of Hollywood, 41
Free Trade Area of the Americas, 62
Friendster, 35
Fryer, Bronwyn, 184
Fulkerson, Richard, 74, 75, 90

Gabilondo, Joseba, 44, 46, 48, 56
Gaia Online, 4, 15-16
Gallagher, Susan E., 140, 143; *Don't Look Now*, 140
Games, David, 133
Games, Marcelo, 133
Garnham, Nicholas, 67, 90
Garofalo, Gitana, 181, 185
Garrison, Ednie, 166, 184
Geekgirl, 179
Genet, Jean, 150
genocide, 18
George, Diana, 62, 79, 87, 90, 92
Gere, Anne Ruggles, 169, 184; *Intimate Practices*, 169
Gerrard, Lisa, 182, 184
Gerz, Jochen, 202
Gianotto, Alison, 177
Gibson, William, 11, 21

Gigliotti, Carol, 10, 21
Giovanni, Nikki, 177
Giroux, Henry, 84, 182
Gleick, James, 14, 21
Gods and Generals, 135
Goggin, Maureen, 63, 90
González, Jennifer, 79, 81, 90
Goodman, Stephen, 62, 82, 90;
  Teaching Youth Media, 82
Google, 134, 215
Gorgias, 148
Governor's Prayer Breakfast, 18
Grabill, Jeffery, 47, 56, 89
Graff, Gerald, 74, 90, 140, 143
Graham, Margaret Baker, 73, 90
Granovetter, Mark, 209
Greece, 146
Greeks, 154
Green, Karen, 167, 181, 184
Greenspan, Alan, 112, 121
Gregory, Marshall W., 79, 88
Grimaldi, William, 8
Gromala, Diana, 9, 21
Groner, Rachael, 182
Guare, John, 207, 221
Guattari, Felix, 21, 35, 39, 148, 189;
  Thousand Plateaus, 35
Guilbaud, G.T., 20, 21
Guyer, Carolyn, 15, 21

Haas, Christina, 38, 39, 90
Habermas, Jürgen, 64, 66, 74, 79
Haefner, Joel, 45, 47, 56
Handelman, Don, 202, 204
Hands, Joss, 62, 91
Hanna, Kathleen, 170, 172, 184
Hanson, Susan, 175, 184
Haraway, Donna, 3, 18, 21, 43, 56, 95, 121
Harburg, Germany, 202
Harlem Renaissance, 139
Harmon, Amy 142, 143
Harré, Rom, 143
Harris, Eric, 203
Harris, Roma, 93
Harrison, Bennett, 106, 121; *Lean and Mean*, 106
Harrison, Stanley, ix, 42, 56

Hart, Joseph, 219, 222
Hart-Davidson, Bill, 182
Harvard University, 41
Harvey, David, 105, 121, 142–43;
  The Condition of Postmodernity, 105
Hawhee, Debra, 148, 161
Hawisher, Gail, 62, 81, 91, 181, 182, 184
Hawk, Byron, x
Hayles, N. Katherine, 11, 21, 34, 38–39
Hebdige, Dick, 75, 81, 91; *Subculture: The Meaning of Style*, 81
Hegel, Georg Wilhelm Friedrich, 150
Heidegger, Martin, 8, 158
Hellfire, 172, 184
Heracles, 148, 151
Heraclitus, 153
Hess, Douglas, 62, 91
Hewlett, Jamie, 185
Hill, Charles A., 91
Hill, Elaine N., 51, 57, 78
Hindemann, Jane E., 38, 39
Hine, Lewis, 65, 91
Hironaka, Lani, 110
Hirschkop, Ken, 108–09, 121
Holocaust Museum, 187
HOPE Columbine Atrium and Library Fund, 200
Horner, Bruce, 169, 185
Hossfeld, Karen, 112
Huberman, Bernardo, 220, 221
Hunter, Henry Julian, 102
Hussein, Saddam, 18
Huyssen, Andreas, 204

Indra's Net, 4
International Monetary Fund, 110
Internet Explorer, 126
interpellation, 132
Iraq, 142
Isocrates, 19
Ivie, Robert, 74, 91

JAC, viii, ix, 126, 182
Japan, 142
Jarratt, Susan, 74, 91
Jefferson County School District, 197

Jenkins, Henry, 91
JenniCam, 33
Johnson-Eilola, Johndan, 4, 17, 22, 42, 45–47, 56, 182
Johnston, John, x
Journet, Debra, 85, 91
Joyce, Michael, 3

kairos, 70, 74, 85, 150–52
Kameen, Paul, 157, 159, 161
Kang, Jerry, 22
Karp, Michelle, 181, 185; *Bust Guide to the New Girl Order*, 181
Kelly, Kevin, 219, 221
Kennedy, George, 5, 8, 11
Kennedy, Pagan, 174, 185
Kenny, Anthony, 144
Kent, Thomas, 158, 161; *Post-Process Theory*, 158
Kerford, G.B., 153, 161
Kevin Bacon Game, 207
Kienzler, Donna, 78, 91
Kinder, Marsha, 141; "Labyrinth Project," 141
King, Lesley, 221
Kinneavy, James, 70, 91, 159
Klebold, Dylan, 203
Klein, Calvin, 80
Kline, Nathan S., 42, 56
Kodak, 65–66, 87
Konigsberg, Germany, 208
Kreis, Jean, 120
Kress, Gunther, 62, 70, 73–74, 87, 91
Krugman, Paul, 113

Labalme, Fen, 125
"Labyrinth Project," 141
Laclau, Ernesto, 74, 91
Lamb, Roger, 143
LambdaMOO, 11, 12
Langdale, Cumbria (UK), 221
Langstraat, Lisa, 182
Lanham, Richard, 67, 74, 92, 138, 144, 153, 161
Lauer, Janice, 147, 160
Leave it to Beaver, 138
LeCourt, Donna, 176–77, 185
Lenin, Vladimir, 104, 121
Lévy, Pierre, 3, 19, 22

Lewis, C.S., *ix*
Life, 129
LimeWire, 72
Lin, Maya, 190
Linenthal, Edward, 190, 204
LinguaMOO, 126
Lisa.bitch/dyke/whore zine, 185
Littleton, Colorado, 196, 199
"Living on the Bottom of Silicon Valley," 121
Local Area Networks, 31, 141
logos, 8, 72
Love Bug, 216
luddites, 95
Lunenfeld, Peter, 14, 22
Lupton, Ellen, 28, 39
Lynch, George, 74, 79, 92
Lyon, Arabella, 86, 92

Mac Expo 2005, 92
Macromedia Flash and Director, 139
Macrorie, Ken, 74, 92
Macy Conferences, 6
Magnuson, Ann, 165, 185
Mailloux, Steven, 119, 121
Malinowski, Bronislaw, 6
Manhattan, New York, 188, 202
Mann, Steve, 42, 56
Manovich, Lev, 25, 26, 39; *The Language of New Media*, 25
Manpower, 111
Maoism, 119
Martin, Alan, 185
Marx, Karl, 48, 57, 73, 96–103, 110–12, 115, 119–21; *Capital*, 102, 110; *The Communist Manifesto*, 96, 100; *The German Ideology*, 48
Marxism, 106, 119, 140
Marxist critique, ix
Marxist idealism, 95
Massumi, Brian, 26, 36, 37, 39, 155, 156, 159–61; *Parables for the Virtual*, 36
Matrix Reloaded, The, 126
Mauser, Daniel, 200, 201, 203
Mavor, Anne S., vii, 141–43
MaximumRocknRoll, 167
Maybury, Mark T., 87, 92

McAllister, Ken, ix, 95, 120, 122, 142, 144
McChesney, Robert Waterman, 61, 92, 98, 122
McComiskey, Bruce, 64, 73, 79, 80, 92
McDonough, Jerome P., 22
McLaughlin, Becky, 182
McWilliams, Peter, 192; *How to Survive the Loss of a Love*, 192
Mediterranean Basin, 126
Melksins, Peter, 106, 107, 109, 122
Memmott, Talan, 34, 38; *From Lexia to Perplexia*, 34
Menchaca, David, 142, 144
Menzies, Heather, 109, 110; *Whose Brave New World?*, 109
Messaris, Paul, 76, 92
Metzger, David, 15, 22
Miami, Florida 62
Michel, Anthony, ix
Michigan State University, 88
Microsoft, 140
Microsoft Word, 3
Milgram, Stanley, 207, 208, 209, 221
Miller, Carolyn, 151, 152, 161
Minority Report, 126
MIT, 6, 55, 125, 126, 128
Mitchum, Robert, 128
MMC Technology, 112
Moje, Elizabeth, 87, 94
Monroe, Barbara, 66, 67, 69, 92
Monument Against Fascism, 202
MOOs, 8, 20, 55, 126, 127
Moran, Charles, 96, 97, 101, 105, 112–15, 122
Moriarty, Tom, 182
Mosco, Vincent, 98, 99, 103, 122
Mouffe, Chantal, 64, 74, 91, 92
MP3 player, 31
Ms. Magazine, 180
MSN, 220
MUD, 8, 127
multiliteracies, 85
Murphy, James, 22
Museum of Victoria, Australia, 178
Myers, Mike, 207

Nair, Yasmin, 182
Napoleon, 18
Nation, The, 129
National Aeronautics and Space Administration, 141
National Research Council, 141
National Rifle Association, 201
National Science Foundation, 141
National Security Agency, 141
Native Americans, 18
Nature, 212
Nelson, Cary, 120, 122
Netscape, 126
network theory, 207–222
Neuromancer, 11
Neuwirth, Christine M., 90
New Dimensions, *viii*
New Guinea, 6
New London Group, 87, 92
New Museum of Contemporary Art, 178, 179
New York City, 82, 199, 217
New York Times, 142, 181
NewsPeek, 125, 126, 127, 128, 136
Newton, Isaac, 151
New Zealand, 217
Nguyen, Mimi, 167, 170, 172–73, 179, 185
Nietzsche, Friedrich, 158
Night of the Hunter, The, 128
Nike, 118
Nimmo, Beth, 200
Nora, Pierre, 187, 188, 202, 204
Norgaard, Rolf, 84, 85, 92
North Carolina, 203

O., Cindy, 185
Occupational Safety and Health Administration, 102
Ohmann, Richard, 104, 122; *Selling Culture*, 104
Oklahoma City Memorials, 188, 190
Olson, Gary A., *vii*
Opera, 126
Orkut, 35
Oxford English Dictionary, 72

Page, Barbara, 16, 22
Paine, Thomas, 165

Pastor-Satorras, Romualdo, 221, 222
pathos, 8, 11, 72, 146, 155
Payne, Michelle, 171, 185
Peel, Frank, 95, 122
Peeples, Tim, 182
Pendakur, Manjunath, 93
Penguin Press, 181
Pentagon, 187
Pentium I, 50
persuasion, 61–94
Peterson, Helana, 178
Petrovich, Lucy, 141
phantasization, 138
Phelan, John, 64
phronesis, 7, 11
pisteis, 11
pistis, 8
Pixar Animation, 9
Planet Grrl, 185
Plato, 154
Pogue, David, 142, 144
Poisson distribution, 210
Polsky, Allyson, 31, 39
Pope, Richard W., 201
Porter, David, 204
Porter, James E., 19, 22, 189; *Rhetorical Ethics and Internetworked Writing*, 19
Poster, Mark, 204
"Post-feminism Sell Sell Sell," 185
Postman, Neil, 64
Poulakos, John, 148, 161
PowerPoint, 52
Powers, Ann, 185
Pratt, Geraldine, 184
Pratt, Mary Louise, 84, 93, 175
praxis, 98
Pray for President Bush, 205
Pregal River, 208
Pre-Socratics, 153
"Prisoner's Dilemma," 213
Prodicus, 148
public sphere, 64, 65

Racheljoyscott.com, 205
Rachel's Tears: The Spiritual Journey of Columbine Martyr Rachel Scott, 200
racism, 18

Ramona, 218
Rand Corporation, 209
Ray, Thomas, 218, 219, 222
Red Cross, American, 192
Reeve, C.D.C., 8, 22
Renyi, Alfred, 208, 210, 212, 220
Revolutionary Worker, The, 110,
Reynolds, Patrick, 222
Rheingold, Howard, 35, 36, 39
rhetoric, 8, 20; deliberative, 20; *epideictic*, 20; judicial, 20
Ridolfo, Jim, ix, 62, 71, 83
Ringley, Jennifer, 33
Riot Grrrl, 5, 167, 170, 185
Robertson, Pamela, 172, 185
Robins, Kevin, 120, 122
Rorty, Amelie, 159, 161
Rosenberg, Jessica, 181, 185
Rosenfield, Lawrence William 19, 22
Ruby, Jay, 76–78, 93
Ruggill, Judd Ethan, 133, 142, 144
Russell, David R., 88
Ryan, Marie-Laure, 22

SAFE Colorado, 200
Said, Edward, 142, 144
Sale, Kirkpatrick, 95, 122
Salkowski, Joe, 178, 186
Sandata, Sabrina, 172, 173, 186
San Francisco State University, 112
San Jose, California, 112
Sandia National Laboratory, 141
Santa Clara County Center for Occupational Safety and Health, 110
Santa Fe Institute, 214, 220
Sapir, Edward, 79
Sausalito, California, 195
Scandinavia, 142
Schiller, Dan, 120, 122
Science, 214
Scott, Darrell, 198, 203
Scottish philosophers, 98
Selber, Stuart A., 93
Selfe, Cynthia, 45, 47, 57, 81, 89, 96–97, 101, 105, 112–15, 123, 182, 184; "Technology and Literacy," 46

Selfe, Richard, 45, 47, 57
Selzer, Jack, 39, 73, 93
Semiotics and the Media website, 82
Senft, Theresa, 38
Shalev-Gerz, Esther, 202
Shalizi, Cosma, 220, 222
Shamgar-Handelman, Lea, 202
Shapiro, Michael, 77, 93
Sheridan, David, *ix*, 62, 82, 93
Sherman, Aliza, 186
Shipka, Jody 62, 71–72, 87, 93
Shirky, Clay, 35, 39; "Social Software," 35
Signs, 181
Silberman, Seth Clark, 181, 183
"Silicon Nightmares," 123
Silicon Valley, 106, 110–12
Slant, 172
Slater, Don, 65–67, 69, 93
Smith, Adam, 98–99, 123
Smith, Alvy Ray, 9, 22
Socolovsky, Maya, *x*
Somerville, Siobhan, 182
Sontag, Susan, 76, 93
sophistic tradion, 19, 80
sophrosyne, 159
Sosnoski, James, *ix*, 142, 144
SoundWriting, 42
Soviet Union, 209
Spin, 181
Spivak, Gayatri, 55, 142, 144, 150
Spy Who Shagged Me, The, 207
St. Paul, Minnesota, 217
Stoller, Debbie, 180; *Bust Guide to the New Girl Order*, 181
Storyspace, 4, 15, 16
Stream, Carol, 197, 205
Streeter, Thomas, 82, 93
Strickland, Stephanie, 221–22
Strogatz, Steven, 212–14, 220, 222
Stroupe, Craig, 62, 93
Structured Programming protocol, 45
Students Against Sweatshops, 118
subaltern counterpublics, 173
Sullivan, Laura, 51, 56
Sullivan, Patricia, 81, 91, 181–82, 184

Summerstein, Evelyn, 182, 186
Syverson, Margaret, 158, 160–61

360degrees, 77, 78
Tagg, John, 65, 67, 69, 93
Tan, Ed, 93
Tannen, Deborah, 74, 93
Taormino, Tristan, 167, 181, 184
Taylor, Charles, 51, 150–55, 157, 161
Taylor, Mark C., 25, 26, 40, 145, 148, 149, 156, 161; *Hiding*, 27, 28, 29, 31; *Moment of Complexity*, 25, 26, 28, 155
Taylor, Todd, 27, 57
techne, 86
Tennyson, Alfred, 126
terministic screens, 38
terrorism, 18
Third World, 46, 51, 100, 115
Thorburn, David, 91
Thuror, Lester, 113
Tierra, 218, 219
Time, 129, 180
TIVO, 135
Tjaden, Brett, 220, 222
Trimbur, John, 62, 68, 73, 85, 87, 88, 94, 120, 123
Trobiand Islanders, 6
Trumpener, Katie, 202, 205
Turkle, Sherry, 9, 23
tyche, 11

Ulmer, Greg, 142, 144, 147, 161; *Heuretics: The Logic of Invention*, 147; *Teletheory*, 142
United States Department of Commerce National Telecommunications and Information Administration, 123
United States Holocaust Memorial Museum, 190
United States Senate, 201
University of Phoenix, 117
University of South Wales, 179

Vale, V., 167, 174, 186
Van Leeuwen, Theo, 62, 70, 73, 74, 87, 91

Vatz, Richard, 149, 150, 162
Venn diagrams, 63
Vespignani, Alessandro, 222
Vietnam Veterans Memorial, 187, 190
virtual rape, 12
Virtualmemorials.com, 205
Vitanza, Victor, 153, 162; *Negation, Subjectivity, and the History of Rhetoric*, 153

Wade, Suzanne, 87, 94
Wagner, Robert, 207
Wagner-Pacifici, Robin, 205
Wahlster, Wolfgang, 87, 92
Walker, Alice; *Color Purple, The*, 137
Wallace, Susan, 198, 205
Walthrop, M. Mitchell, 220, 222
Ward, Irene, 62, 94
Warnick, Barbara, 19, 20, 23
Wasson, Glen, 220, 222
Watt, Stephen, 120, 122
Watts, Duncan, 212–14, 220, 222
Webster, Frank, 120, 122
Weisser, Christian R., 94
Welch, Kathleen, 19, 23, 62, 94
Wellington, Jan, 182
Welsh, Scott, 64, 80–82, 94
West, D.B., 222
West, Tom, 182
Wettach, Gabriel, 182
Whitaker, Elaine E. 51, 57
White, E.C., 151; *Kaironomia*, 151
White, Michele, 33, 34, 38, 40; "Too Close to See," 34, 38
Whitewater, Colorado, 201
Wiener, Norbert, 6, 10, 14, 23; The *Human Use of Human Beings*, 14
Wild Things, 207
Williams, Raymond, 80, 94
Windows 3.1, 43
Windows 95, 43
Windows 98, 43, 50
Windows ME, 43, 50
Winfrey, Oprah, 137
Wired magazine, 138
WorldPeek, 139
World Trade Center, 187, 188
World Trade Organization, 36, 110

World War I, 103
World War II, 113, 119
Worsham, Lynn, *viii*
Wyard, Peter, 87, 94
Wysocki, Anne, 73, 74, 87, 88, 94

Yahoo!, 220
Young, Iris Marion, 27, 40, 74, 94
Young, James, 189, 205

Zerbe, Mike, 182
zine culture, x
Žižek, Slavoj, 193, 194, 205
Zoba, Wendy Murray, 197, 198, 203, 205

Printed in the United States
201645BV00004B/277-345/P